T0327563

Wireless LAN Radios

IEEE Press Series on Digital & Mobile Communication

The IEEE Press Digital and Mobile Communication Series is written for research and development engineers and graduate students in communication engineering. The burgeoning wireless and personal communication fields receive special emphasis. Books are of two types, graduate texts and the latest monographs about theory and practice.

John B. Anderson, *Series Editor*
Ericsson Professor of Digital Communication
Lund University, Sweden

Advisory Board

John B. Anderson
Dept. of Information Technology
Lund University, Sweden

Joachim Hagenauer
Dept. of Communications Engineering
Technical University
Munich, Germany

Rolf Johannesson
Dept. of Information Technology
Lund University, Sweden

Norman Beaulieu
Dept. of Electrical and Computer
Engineering,
University of Alberta,
Edmonton, Alberta, Canada

Books in the IEEE Press Series on Digital & Mobile Communication

John B. Anderson, *Digital Transmission Engineering, Second Edition*

Rolf Johannesson and Kamil Sh. Zigangirov, *Fundamentals of Convolutional Coding*

Raj Pandya, *Mobile and Personal Communication Systems and Services*

Lajos Hanzo, P. J. Cherriman, and J. Streit, *Video Compression & Communications over Wireless Channels: Second to Third Generation Systems and Beyond*

Lajos Hanzo, F. Clare, A. Somerville and Jason P. Woodard, *Voice Compression and Communications: Principles and Applications for Fixed and Wireless Channels*

Mansoor Shafi, Shigeaki Ogose and Takeshi Hattori (Editors), *Wireless Communications in the 21st Century*

Raj Pandya, *Introduction to WLLs: Application and Development for Fixed or Broadband Services*

Christian Schlegel and Lance Perez, *Trellis and Turbo Coding*

Kamil Zigangirov, *Theory of Code Divison Multiple Access Communication*

Arya Behzad, *Wireless LAN Radios: System Definition to Transistor Design*

Wireless LAN Radios
System Definition to Transistor Design

Arya Behzad

IEEE SERIES ON
**DIGITAL
& MOBILE**
COMMUNICATION

John B. Anderson, *Series Editor*

IEEE Press Series on Microelectronic Systems
Stuart K. Tewksbury and Joe E. Brewer, *Series Editors*

IEEE Solid-State Circuits Society, *Sponsor*

IEEE PRESS

WILEY-INTERSCIENCE
A JOHN WILEY & SONS, INC., PUBLICATION

Contents

Preface

This book provides a high-level overview of the design of radios for wireless local area network (WLAN) systems. In doing so, it spends a considerable amount of time describing the unique aspects of the WLAN system. It is important to understand these unique aspects in order to be able to design an optimal radio for this system. Only with proper high-level system understanding will a designer be able to trade off the ever-present challenges that are to be made.

As a high level and concise overview, this book does not discuss in detail any of the aspects covered. However, it enables the reader to grasp a good understanding of the overall challenges faced in the design of radios for WLAN systems.

The book covers a variety of topics, from communication system concepts to transistor level circuit implementations and trade-offs. Therefore depending on the reader's area of expertise, he or she may find certain chapters easier to follow than others. However, a system designer, for example, should be able to have a good understanding of the challenges faced by the circuit designer. Similarly, this book enables a circuit designer to be able to comprehend the reasoning behind the block specifications that the system designer has passed on to him or her. Given that current and future generation radios will require more and more system level calibrations, such an understanding on both sides is essential to designing the next generation radios for WLAN applications.

This book is organized as follows. A quick introduction is presented in this preface. Chapter 1 describes the various flavors of the 802.11 PHY standard and the system and radio requirements associated with these PHY standards. Various receiver and transmitter architectures that can be utilized in designing WLAN systems is described in Chapter 2 and the various trade-offs associated with these architectures are described. Chapter 3 outlines in fairly significant detail the analog impairments and issues associated with implementing WLAN radios. Chapter 4 discusses transistor level implementation of some key radio building blocks. Chapter 5 discusses several calibration techniques used in the design of WLAN radios. In Chapter 6, two case studies are presented, one of the design of a full 802.11a WLAN radio and another of a calibrated transmitter for a WLAN application. Finally, a brief conclusion is presented.

Upon completing the study of this book, the reader should have a strong high-level overview of the multitude of trade-offs that can be made in the design of radios for the various flavors of WLAN systems. The trade-offs made are a result of the complex interactions of the choice of radio architecture, the choice of process technology, the choice of the calibration algorithms utilized, and several other factors.

I acknowledge my colleagues at Broadcom for their contributions to the many WLAN chips that have been discussed and referenced in this book, including the folks on the RF design team, RF layout team, systems design team, operations team, and central engineering team. I also thank Broadcom management for supporting and authorizing the publication of this book. In addition to the referenced published material in the book, some of the figures in this book are extracted from various presentations. I thank the authors of these presentations: Rohit Gaikwad, Antonio Montalvo, David Su, Jason Trachewsky, Tyson Tuttle, and Iason Vassiliou. I thank Klaas Bult for his review of the book. Finally, many thanks to the staff at IEEE Press and Wiley for their work on the manuscript.

And, of course, my great gratitude goes back to my parents and brother for their lifelong love and support, and to my wife and children for their love and for putting up with me and my schedule while I worked on this book.

ARYA BEHZAD

San Diego, California
September, 2007

Acronyms

AACI	Alternate adjacent channel interferer
ACI	Adjacent channel interferer
AD	Amplitude distortion
ADC	Analog to digital converter
AFC	Automatic frequency correction
AM	Amplitude modulation
AP	Access point
ASP	Average selling price
AWGN	Additive white Gaussian noise
BALUN	Balanced-unbalanced (a single-ended to differential converter)
BPF	Bandpass filter
BPF	Bandpass filter
BT	Bluetooth
BW	Bandwidth
CCK	Complementary code keying
CMOS	Complementary metal-oxide semiconductor
CW	Continuous wave
DAC	Digital-to-analog converter
dBc	Decibels relative to the carrier
DFT	Discrete Fourier transform
DSB	Double sideband
DSL	Digital subscriber line
DSP	Digital signal processor
DSSS	Direct sequence spread spectrum
DUT	Device under test
DVB	Digital video broadcasting
EDGE	Enhanced data rate for GSM evolution
EVM	Error vector magnitude
EWC	Enhanced wireless consortium
FCC	Federal Communications Commission
FHSS	Frequency hopping spread spectrum
FM	Frequency modulation

fpBGA	Fine-pitch ball grid array
FSK	Frequency shift keying
GI	Guard interval
GSM	Global system for mobile communications
HBT	Heterojunction bipolar transistor
HD	Harmonic distortion
HPF	High-pass filter
HPVGA	High-pass variable gain amplifier
IDFT	Inverse discrete Fourier transform
IF	Intermediate frequency
IMD	Intermodulation distortion
IO	Input/output
IP	Internet protocol
IP2	Second-order intermodulation product
IP3	Third-order intermodulation product
I/Q	In-phase/Quadrature phase
ISI	Intersymbol interference
ISM	Industrial—scientific—medical
LINC	Linear amplification using nonlinear components
LNA	Low-noise amplifier
LO	Local oscillator
LOFT	Local oscillator feed-through
LOS	Line of sight
LPCC	Leadless package chip carrier
LPF	Low-pass filter
MAC	Media access control
MCS	Modulation and coding scheme
MEMS	Micro-electro mechanical systems
MIMO	Multi-in multi-out
MIMO	Multi-in multi-out
MISO	Multi-in single-out
NLOS	Nonline of sight
OFDM	Orthogonal frequency division multiplexing
OOB	Out of band
P1dB	1-dB compression point
PA	Power amplifier
PAD	Power amplifier driver
PAPR	Peak-to-average power ratio
PAR	Peak to-average ratio
PCB	Printed circuit board
PD	Phase detector

PD	Phase distortion
PDF	Probability distribution function
PFD	Phase-frequency detector
PGA	Programmable gain amplifier
PHEMT	Pseudomorphic high electron mobility transistor
PHY	Physical layer
PLCP	Physical layer convergence protocol
PLL	Phase locked loop
PPDU	PLCP protocol data unit
PPM	Parts per million
PTAT	Proportional-to-absolute temperature
QAM	Quadrature amplitude modulation
QoS	Quality of service
RMS	Root mean square
RSSI	Received signal strength indicator
Rx	Receiver
SAW	Surface acoustic wave
SDM	Spatial division multiplexing
SiGe	Silicon germanium
SIMO	Single-in multi-out
SNDR	Signal-to-noise plus distortion ratio
SNR	Signal-to-noise ratio
SSB	Single sideband
STA	Station
TCP	Transmission control protocol
TIA	Transimpedance amplifier
TR	Transmit—receive
TSSI	Transmit signal strength indicator
Tx	Transmitter
UNII	Unlicensed national infrastructure for information
VCO	Voltage-controlled oscillator
VNA	Vector network analyzer
VOIP	Voice over internet protocol
VSA	Vector signal analyzer
WCDMA	Wideband code division multiple access
WEP	WLAN encryption protocol
WLAN	Wireless local area network
XO	Crystal oscillator
ZIF	Zero intermediate frequency

CHAPTER 1

802.11 Flavors and System Requirements

1.1 DEFINITION

What is a wireless local area network (WLAN)? A WLAN system, shown in its most general form in Figure 1.1, consists of a network hardware backbone, along with a series of detached components. These detached components may include computer desktops, computer laptops, personal digital assistants (PDAs), cell phones, gaming systems, security cameras, printers, and appliances as clients. Using radio-frequency (RF) technology,[1] the WLAN system would then allow the clients to access local area network resources while physically being detached from this network. At the same time, the clients are capable of communicating with one another (typically indirectly and through access points rather than peer-to-peer networks) while physically being detached from one another. A WLAN system can transmit data, video, and/or audio.

A WLAN system may be deployed as a stand-alone network or in tandem with a wired network. As compared to a wired network, a WLAN system offers several advantages and suffers some disadvantages.

On the positive side, a WLAN system allows mobility and flexibility. For existing infrastructures, especially those with high user density (hotel rooms, apartment complexes, etc.), it offers the lowest cost and most flexible method of connectivity. Whereas it may be inexpensive to install category 5 (CAT5) wiring for new buildings, to do so in an existing building is quite costly and inconvenient. Given the cost of WLAN chipsets at the current time, it would be much more cost effective to install a simple WLAN system than to run wires through such structures. At the same time, even if CAT5 wiring is installed, for example, in every room in a newly constructed home, it is often not exactly "at the right place." Wireless LAN would offer

---

[1]Of course an alternative wireless technology such as infrared signaling may be used, but the most common WLAN systems today utilize RF technology. As a result the term "WLAN" is almost exclusively utilized to refer to WLAN communications utilizing RF technology.

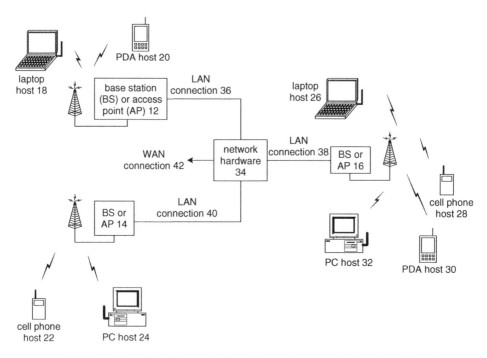

Figure 1.1 Example of WLAN network displaying various associated nodes and backbone network.

the flexibility of connectivity anywhere in the home, without an a priori requirement to determine the precise locations of the network taps.

On the other hand, a WLAN system is typically never as secure as a dedicated (for example, T1) or even shared (for example, cable modem) wire connection. The mere fact that the medium is shared by potentially many users and no physical connection is required to "tap" into the network makes the WLAN network more susceptible to "hacking" and "spoofing." At the same time, various research studies have shown that many WLAN users fail to properly activate the proper encryption options on their access points and thereby make themselves susceptible to hackers.

Recent developments in encryption technology and standards as well as recent software drivers that simplify the installation process of a protected WLAN clients and access points, however, have significantly improved the situation as compared to the early days of WLAN history.

In terms of communication speed, also, WLAN networks are typically a generation or so behind their wired LAN counterparts. This is due to the difficulties associated with the medium of communication (air). For example, in the indoor environment these challenges include propagation losses through the air medium and through walls, multipath caused by reflections from objects and people, and interference due to other wireless communication devices and interferers such as microwave ovens.

It should have become apparent by now that neither a wireless network nor a wired network is capable of providing all the desired characteristics and amenities. Quite often, therefore, an "optimal" network is one that is constructed of a wired LAN "backbone" and is complemented by a WLAN network that would provide flexibility and reconfigurability.

1.2   WLAN MARKET TRENDS

We will spend a few paragraphs discussing the WLAN market trends. The objective here is to put into perspective the phenomenal growth this market has experienced while emphasizing the extremely competitive nature of this market. Thousands of pages of analyst reports are published annually on this subject and we will make no attempt to cover the details that are covered in such reports. Further the WLAN market conditions are quite fluid and change almost quarterly, and therefore the absolute numbers (and possibly even trends) may not hold in the future.[2]

Wireless LAN has been one of the fastest growing segments of the semiconductor market. Despite the slow sales growth (or even decline) of semiconductors for the early 2000 years the WLAN chipset market has grown quite significantly in those years. As seen in Figure 1.2a, the number of WLAN users has grown quite rapidly, especially in the home market. The enterprise has been growing fairly significantly but not nearly as quickly as the home market. The primary reason for this is the concern of the enterprise customer about security. In the early days of WLAN, a major news item about a few University of California—Berkeley Computer Science students breaking the fairly vulnerable 48-bit encrypted WLAN encryption protocol (WEP) did not help the confidence level of the enterprise customers either. By using 128-bit encryption and further enhancements to the security protocols, those issues have been addressed by the standard now (more on this topic later). Of course, the encryption techniques will be continuously updated and strengthened as issues are discovered and as the hackers improve the sophistication of their techniques.

Quality of service (QOS) has also been an issue that has held back the adoption of WLAN by the enterprise as well as certain home users. Certain WLAN applications require a guaranteed maximum latency and would need

---

[2]Unlike, hopefully, the technical discussions in this book which should hold "forever"!

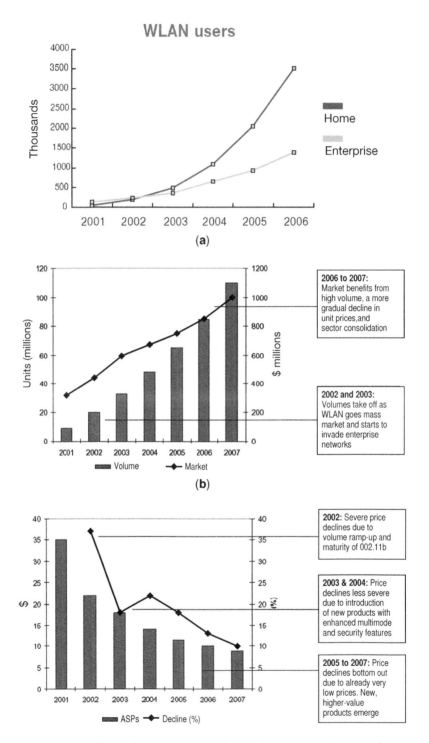

Figure 1.2 (**a**) WLAN growth trend in home and enterprise markets, (**b**) WLAN chipset volume growth chart, and (**c**) historical decline trend in chipset average selling price. (*Sources:* lightreading.com, newsweek.com.)

to be prioritized over other types of network traffic. An example of such la-tency-sensitive packets is voice-over-Internet protocol (VOIP) packets. VOIP is the standard used to do telephony over an Internet protocol (IP)–based wired or wireless LAN. The resolution and proper implementation of QOS on the WLAN networks would therefore accelerate the adoption and sale of WLAN devices.

Figure 1.2b shows the growth of the 802.11 chipset volumes and the market values extrapolated to the year 2007. The rapid growth of chipset volumes is apparent in this figure and at first may look like an extraordinary business opportunity! However, before trying to put a startup company together to address this market, one needs to review Figure 1.2c. This chart shows the rapid decline in the average selling prices of the chipsets caused by the increase in volume. This steep price drop can be attributed to many factors, such as increase in the selling volumes, very high levels of integration, the numerous players in the market and the resultant competitive nature of the business. In the past few years, the extreme competitive nature of the business has caused many of the smaller and some of the larger players to exit the market segment all together.

Figure 1.2c shows how the average selling prices (ASPs) have dropped very quickly early on as the volumes were ramping up. This period was followed by some price stabilization and then further reduction in prices. The stabilization points correspond to times in the market in which the chipset vendors started offering new features and were therefore able to demand higher prices. This phenomenon temporarily reduces the erosion of price in the WLAN chipset market. For example, in 2003 the steep decline in prices was slowed by the introduction of the 802.11g standard, which allowed for much higher data rates than the traditional 802.11b standard.

Of course, eventually prices will continue their downward trend. It is therefore critical for the chipset industry to keep on innovating and offering newer features. This is necessary in order to be able to offer newer higher margin products as the older ones become commodity items and decline in their profit margins.

A factor that can affect and slow down the reduction in the average selling prices is the addition of new features and new building blocks within the chipsets. So the addition of such blocks into the chips allows the manufacturers of the chips to demand higher prices at the same time the end customer would have a lower bill-of-materials cost.

In summary, this steep price decline and the extreme competitive nature of the WLAN chipset market dictate one of the most important WLAN chipset design requirements: design for low cost. Design for low cost, in turn, translates into design in the lowest possible cost technology, highest levels of integration, smallest possible die size, low packaging and testing cost, and high yields. Since not all of these criteria can be simultaneously satisfied, designers will have to make complex trade-offs to come up with the lowest possible final product cost. Combined with other product requirements such as time to market and system performance, the designers are required to make many difficult choices early on in the design that could quite likely result in a product being successful or a dud.

These trade-offs will be discussed in much more detail in the subsequent chapters.

There are various WLAN standards, such as HyperLAN and the Institute of Electrical and Electronics Engineers (IEEE) 802.11, but at this time, in the United States, Europe, the Far East, as well as elsewhere in world, the 802.11 standard has become the standard of choice for WLAN and will therefore be emphasized in this book.

1.3   HISTORY OF 802.11

In 1990, the IEEE 802 executive committee established the 802.11 working group to create a WLAN standard. The standard specified an operating frequency in the 2.4-GHz ISM (industrial, scientific, and medical) band and began laying the groundwork for a cutting-edge technology. After seven years, in 1997, the group approved IEEE 802.11 as the world's first WLAN standard with data rates of 1 and 2 Mbps. Having great foresight, the executive committee predicted the need for a more robust and faster technology. Therefore, immediately, the committee began work on another 802.11 extension that would satisfy such future demands. Within 24 months, the working group approved two project authorization requests for higher rate physical (PHY) layer extensions to 802.11. The two extensions were designed to work with the existing 802.11 medium access control (MAC) layer, with one being the IEEE 802.11a—5 GHz and the other IEEE 802.11b—2.4 GHz.

The IEEE 802.11 has gained acceptance over competing standards such as HyperLAN and will be the emphasis of this book. The 802.11 is a specific standard that defines the MAC and PHY layers of a WLAN. The original 802.11 standard is a MAC standard plus a low data rate PHY which supports only 1- and 2-Mbps data rates. This first version of the standard operates at the 2.4-GHz ISM band and allows the vendors to choose between a direct sequence spread spectrum (DSSS) and a frequency hopping spread spectrum (FHSS) implementations. As mentioned above, 802.11b is a PHY extension to the original 802.11 standard. It also operates at the 2.40-GHz band and allows for higher data rates of 5.5 and 11 Mbps. It uses a technique known as complementary code keying (CCK).

The 802.11a is another PHY extension to the 802.11 standard. It operates at the 5-GHz unlicensed national infrastructure for information (UNII) band and allows for data rates of 6–54 Mbps. It uses a technique known as orthogonal frequency division multiplexing (OFDM; this technique will be discussed in much more detail in later chapters).

The 802.11g was the next extension to the 802.11 standard. It operates at the 2.4-GHz ISM band and allows for data rates ranging from 1 to 54 Mbps. The 1- and 2-Mbps rates are operated in the DSSS mode whereas the $5\frac{1}{2}$- and 11-Mbps rates are operated in CCK mode. Additionally, rates at 6 to 54 Mbps are operated in OFDM mode. The 802.11g standard borrows the OFDM technique and data rates from the 802.11a standard but operates at the 2.4-GHz ISM band. It can therefore operate at very high data rates while being backward compatible with the 802.11b standard.

In addition to these standards, which have already been approved, the 802.11 committee has "working groups" to evolve and enhance the standard. Here are some examples:

- **802.11e** Tasked to improve QOS. The inclusion of a QOS protocol is essential for tasks that require low latency such as VOIP.
- **802.11i** Tasked to improve encryption. A reliable and hard-to-break encryption technique is essential for the wide adoption of WLAN by the enterprise customer.
- **802.11f** Would allow for an interaccess protocol for easy communication between access points.
- **802.11h** Allows for dynamic frequency selection, and transmit power control. By utilizing dynamic frequency selection, interference between various users would be reduced, and therefore the effective capacity of the cell and therefore the network would increase. Further, by utilizing transmit power control, the minimum required transmit power would be utilized in communication between the access points and the mobile units. This would also reduce cochannel interference and therefore increase the network capacity.
- **802.11n** Allows for multichannel and higher data rate 802.11 in the 2.4- and 5-GHz bands. As of the date of the publication of this book, a "pre-n" standard has been approved by the IEEE, but the final draft has not yet been ratified. The pre-n standard utilizes optional higher order constellations, wider bandwidths, and multi-in, multi-out (MIMO) techniques to dramatically increase the data rate, effective range, and reliability of the WLAN. The 802.11n standard is expected

to be fully backward compatible with the 802.11a and 802.11g standards. We will briefly discuss 802.11n in more detail in Chapter 7.

802.11: b, a, OR g?

The three commonly known versions of the 802.11 PHY are 802.11a, 802.11b, and 802.11g. As described earlier, the 802.11a and 802.11g standards offer much higher speed that 802.11b. However, the advent of 802.11a and g will not necessarily result in the demise of 802.11b in the immediate future. There are applications that would require the lowest power consumption and/or the lowest system cost, and in such cases a stand-alone 802.11b solution may still be the best solution in the immediate future. On the other hand, most system vendors have migrated to 802.11g solutions, which are backward compatible with 802.11b and allow the higher data rates. As the cost of 802.11g solutions drop and their power consumption reduces, this trend will accelerate.

As an alternative to 802.11b and g, if the operator requires a higher data rate, higher user density, and network capacity, he or she would have to choose 802.11a because of the availability of a much wider spectrum at the 5-GHz band and the higher data rates offered by 802.1a.

For longer ranges and higher data rate applications the operator would probably choose 802.11g. The 802.11g offers the added benefit of being backward compatible with 802.11b, which has the largest existing base.

Many applications will probably eventually move to a multiband a/g solution, which would by definition also be backward compatible with 802.11b solutions. This will happen as the cost of multiband solutions drops as a result of further integration and possibly other factors.

Table 1.1 qualitatively shows the advantages and disadvantages of the existing PHY standards. The highlights are listed below.

Currently, there is a much larger existing base for the 802.11b solution. Of course, since 802.11g systems are backward compatible with 802.11b,

Table 1.1 Relative Advantages and Disadvantages of 802.11a, b, and g

Standard	Existing Base	Data Rate	Range	Lack of Interferers	Spectrum Availability	Power Consumption	System Cost
802.11b	++++	+	++++	+	+	++++	++++
802.11a	+	++++	+++	+++	+++	++	++
802.11g	++	++++	++++	+	+	+++	+++

they would be able to take advantage of the 802.11b existing base at lower data rates.

In terms of data rate, the 802.11a and g have an advantage, with rates up to 54 Mbps.

In terms of range of operation, the 802.11b and g have the advantage because they operate at the lower frequency of 2.4 GHz. Since typically propagation losses are lower at lower frequencies, 802.11b and g systems would be able to operate over longer distances as compared to their 802.11a counterpart for a given transmit power and receiver sensitivity. The free-space loss for cases in which the receiver-to-transmitter distance is much larger than the wavelength is given by the relation

$$L = \left(\frac{4\pi d}{\lambda} \right)^2 = \left(\frac{4\pi df}{c} \right)^2$$

where L is the propagation loss, d is the distance between the transmitter and the receiver, λ is the wavelength of the RF signal, f is the frequency of the signal, and c is the speed of light. Antenna gains, absorption losses, reflective losses, and several other factors are not taken into account in the above equation. An indoor environment is much more complex to model or predict than this formula suggests. The interested reader can refer to many publications on this topic.

This simple equation, however, does show the relation between the transmission frequency and the propagation losses. For example, at a distance of 10 m in free space and with the assumptions listed above, a 802.11g system operating at 2.4 GHz would experience 60 dB of propagation attenuation, whereas an 802.11a system operating at 5.8 GHz would experience 68 dB of propagation losses.

The 802.11a has the upper hand when it comes to lack of interferers. This is due to the smaller existing base at the 5-GHz band as well as the wider available spectrum. Additionally, there are far fewer nonwireless LAN systems operating at the 5-GHz band. Such interferers include microwave ovens, security cameras, and cordless phones.

From a spectrum availability point of view, the 802.11a has several hundreds of megahertz of bandwidth available to it (although the exact frequencies would depend on the country of operation). In most countries, on the other hand, there is no more than 100 MHz available for users in the 802.11b or g bands.

From a power consumption point of view, 802.11b would win against the other standards. This is because it utilizes the simplest modulation technique among the three and therefore does not require a high performance radio front end or a sophisticated signal processing baseband. In particular, an 802.11b modulated signal has a small peak to average ratio, and therefore one can use higher efficiency (but lower linearity) power amplifiers on the transmit side.

From a system cost point of view, currently 802.11b offers the lowest system cost. However, the difference in the cost between 802.11g systems and 802.11b systems has been reducing quickly, and today most users are willing to pay the slightly higher cost of an 802.11g system for the significant gains in throughput.

As an interesting marketing point, the number of 802.11g units shipped in The first quarter of 2004 surpassed the shipped 802.11b solutions in that same quarter.

1.5   802.11b STANDARD

As shown in Figure 1.3a, there are a total of 11 designated channels in the 802.11b/g band in the United States. These channels reside in the 2.4-GHz ISM band. However, as shown in Figure 1.3b, there are only three nonoverlapping channels that can operate under the 802.11b/g standard. Within a given cell, if users operate simultaneously on overlapping channels, the interchannel interference would increase, and the overall channel capacity would decrease. The maximum allowed transmit power in the United States for the 802.11b/g standard is 30 dBm or 1 W.[3] This is quite a high transmit power, and most 802.11b/g solutions today operate at significantly lower transmit powers (in the range of 15 to 22 dBm transmit power). This is because the 2.4-GHz ISM band is adjacent to Federal Communications Commission (FCC)–restricted bands. So when operating in the lowest and highest 802.11b/g channels, often the FCC spectral mask requirements associated with these restricted bands is violated before the 802.11b/g mask is violated. Clearly the more stringent of the two masks would set the maximum allowable transmit power.

Worldwide, there are a total of 14 total channels allocated to the 802.11b/g standard operating at the frequency range of 2.40 to 2.58 GHz. The channels are 5 MHz apart. In the United States channels 1, 6, and 11 are typically used to minimize overlap and therefore reduce interference between operating devices. However, as an example, it is possible for a very high power transmitter operating in channel 1 to have an impact on the

---

[3]Note that this is the average maximum transmit power. Due to potential for large peak-to-average ratio in an OFDM signal, for example, the peak instantaneous power can be significantly more than this.

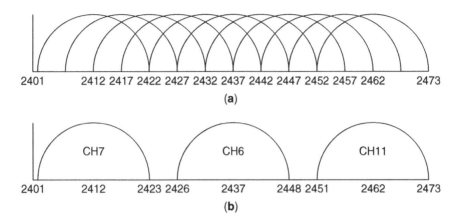

Figure 1.3   IEEE 802.11b/g channel allocations. Note the overlap channels (**a**) as well as the three distinct (nonoverlap) channels (**b**). The x-axis represents frequency in MHz.

throughput of channel 11. Different countries have differing regulations that limit the use of certain channels for 802.11b/g in those countries. For example, in Europe, channels 1 through 13 can be utilized for 802.11b/g operation but at a maximum transmit power of 100 mW. This is done in order to reduce the interference with other ISM band devices.

As mentioned earlier, the original 802.11 standard only allows for 1- and 2-Mbps data rates. In doing so it allows the use of a technique known as DSSS. This technique spreads the data over a wide bandwidth to gain immunity to interferers and multipath reflections. The technique is similar to what is used for the IS-95 cellular code division multiple-access (CDMA) standard.

As an alternative the original standard allows for a FHSS technique. This technique is also designed to improve the immunity of the signal to interferers and multipath channel reflections but, as the name suggests, relies on the carrier frequency to hop around at a pseudorandom center frequency basis. The FHSS technique is similar to what is used in the Bluetooth (BT) standard.

The 802.11b extension to the standard allows for the introduction of higher data rates of 5.5 and 11 Mbps. The 802.11b relies on CCK, a distinct nonsystematic block code which offers both spreading as well as a minimal amount of coding gain. In a sense it can be viewed as a special case of DSSS.

As is typical for any system and any modulation, the signal-to-noise (SNR) requirement for the higher data rates is higher than those for the lower data rates. As such the standard requires a minimum system sensitivity of

−80 dBm for the 1-Mbps data rate and a minimum system sensitivity of −76 dBm for the 11 Mbps. However, today, most systems are capable of delivering much better sensitivity numbers than the standard requires. A state-of-the-art system today can achieve about −98 and −91 dBm "chip sensitivity," respectively, for the 1- and 11-Mbps data rates. The system sensitivity is typically 1 to 2dB worse than the chip sensitivity for the 802.11b operation due to losses of front-end components such as baluns, filters, switches, and board traces at 2.4 GHz.

Table 1.2 summarizes the modulation types and the sensitivity numbers for the various 802.11b data rates.

The 802.11b standard is, in principle and as compared to 802.11g and especially 802.11a, fairly easy to implement. The standard achieves a maximum of 11 Mbps over an equivalent noise bandwidth of 11 to 15 MHz depending on the implementation. This results in a comparatively low spectral efficiency of <1 bit/s/Hz. As a reference, note that a maximum spectral efficiency of > 3 bits/s/Hz is achieved for the 802.11g and 802.11a standards. Of course, in general, wireless communications are limited to much lower spectral efficiencies than those of their wireline counterparts due to the much inferior communication medium (channel). For example, digital subscriber line (DSL) systems, gigabit Ethernet, or cable systems can achieve spectral efficiencies in excess of 10 bits/s/Hz.

Additionally, the 802.1b modulation has a low peak-to-average power ratio (PAPR). This is by no means a constant-envelope modulated signal (like that of Bluetooth, for example), but neither does it have very large PAPR associated with the OFDM coding utilized in the 802.11a and 802.11g standards. The low PAPR characteristic of the 802.11b standard makes the modulation somewhat immune to nonlinearities in the signal path. This characteristic in particular makes the implementation of the power amplifier (PA) in the transmit path much simpler than those required for the 802.11a and g standards.

Table 1.2 IEEE 802.11b/g Allowed Data Rates, Associated Modulation Types, and Required Sensitivities

Data Rate (Mbps)	Modulation	Sensitivity Requirement (dBm)	State-of-the-Art Chip Sensitivity (dBm)
1	D-BPSK	−80	−98
2	D-QPSK		−96
5.5	CCK		−93
11	CCK	−76	−91

Note: Obtained state-of-the-art sensitivity levels are also reported.

1.6   802.11a CHANNEL ALLOCATION

As mentioned earlier the 802.11g channel allocation is identical to that of 802.11b (Fig. 1.3). As such, there are only three nonoverlapping channels available to the users.

One of the advantages of the 802.11a standard as compared to the 802.11g standard becomes apparent in Figure 1.4: There are currently a total of 12 "nonoverlapping" channels available in the United States with proposals at the FCC to open up even more spectrum in the 5-GHz band as part of an expanded unlicensed National Information Infrastructure (NII) spectrum. The large number of channels available in the 802.11a band allow for much higher overall cell and network capacity and less interchannel interference.

As can be seen in Figure 1.5, the statement about the 802.11a channels being nonoverlapping is not completely correct. The spectrum associated with the information content of each channel is designed to be nonoverlapping with its adjacent channels. However, because of imperfect filtering as well as nonlinearities and spectral regrowth in the system, there is a limited amount of spectral leakage from each channel which leaks into its adjacent channels. The magnitude of this leakage is highly regulated by the spectral mask requirements of the standard. The performance of the system in the presence of adjacent channel interferers is also regulated by the standard (more on this later).

In the United States the maximum allowed transmit power for the 802.11a standard is dependent on the subband (Fig. 1.5). In the lower, mid, and higher 802.11a subbands, the maximum transmit power is limited to 16,

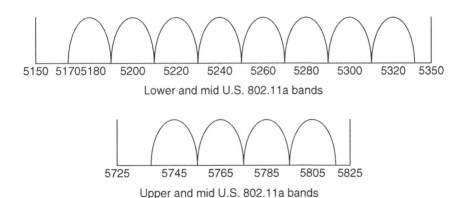

Figure 1.4 Detail of IEEE 802.11a channel allocations in U.S. (total 12 nonoverlapping channels). The lower, mid, and upper bands are shown. Note that no overlapping channels are allowed.

23, and 29 dBm, respectively. The higher subband is primarily intended for long-range outdoor communications.

Various countries allocate different frequency bands for the 802.11a standard. In general, 802.11a systems around the world (non-U.S.) operate in the 4.92- to 5.70-GHz spectrum (Fig. 1.5). Recent proposals have worldwide channels operating as high as 5.845 GHz. For various countries, not only the dedicated frequency channels but also the maximum transmit power per channel as well as various other requirements vary. The interested reader should refer to specific regulations of a given country.

1.7   802.11a AND 802.11g: OFDM MAPPING

The 802.11a and g utilize a technique known as orthogonal frequency division multiplexing, or OFDM. Conceptually, OFDM has been around for a long time. It has been used in a variety of applications for years. These include such applications as digital video broadcasting (DVB) and digital subscriber line (DSL). OFDM does require a significant amount of signal processing horsepower, and such horsepower until recently would consume quite a bit of power consumption. Clearly a high power consumption chipset would not be very suitable for portable applications.

Recent advancements in process technology and also low power design techniques have enabled a dramatic reduction in power consumption of OFDM-based modems. These modems are therefore now suitable for many portable applications such as computer laptops. The push for reducing the power consumption of OFDM-based modems, of course, continues. Further reductions in power consumptions are enabling the integration of WLAN systems in some of the most power-sensitive consumer application gadgets.

OFDM provides a good degree of immunity to multipath fading, which is typically a major problem for high speed wireless communication, especially in an indoor environment. In order to comprehend the concept of multipath fading and its impact on high speed communications in an indoor environment, a brief discussion of the topic is presented in the following section.

1.7.1   Multipath Fading

Multipath propagation, or in short multipath, occurs when signals reflect off of various objects and even people and add constructively or destructively at the receiver antenna. When the signals add destructively, they can significantly impact the quality of the link. This can result in a significant reduction in the throughput of the system. Figure 1.6 depicts multipath when a direct line-of-sight (LOS) path does exist. Figure 1.7 depicts a scenario in

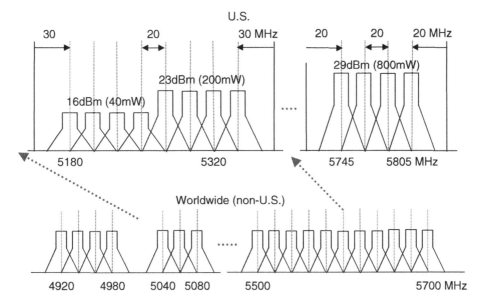

Figure 1.5 Associated power levels for U.S. IEEE 802.11a subbands. The additional world-wide 802.11a subbands are also shown. Note that, although the main channels are nonoverlapping, the channels can interfere with their adjacent channels (as shown) due to inadequate filtering or spectral regrowth.

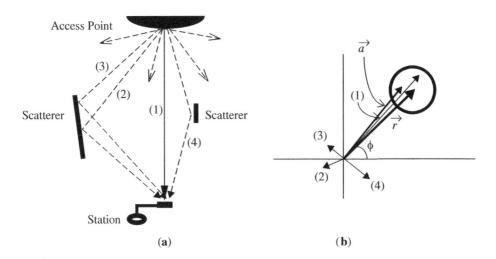

Figure 1.6 (a) Multipath in presence of a line-of-sight signal. (b) Vector space representation.

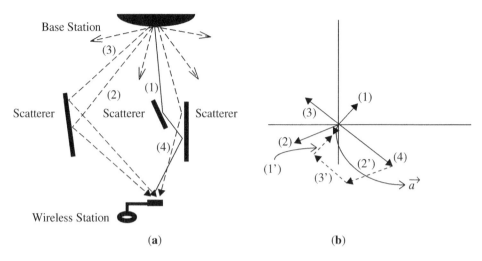

Figure 1.7 (**a**) Multipath response in absence of a LOS signal. (**b**) Vector space representation. Note that the vector magnitudes have been scaled 2 : 1 as compared to Figure 1.6 to simplify visualization.

which a direct LOS does not exist. Clearly, in the latter case, the resultant received signal can be quite small.

Multipath fading is very much environment specific but typically does not exceed about 20 dB in an indoor environment with carrier frequencies in the few GHz range. As described above, multipath is a phenomenon caused by the multiple arrivals of the transmitted signal to the receiver due to reflections off of "scatterers." The gain and phase of these reflections can be modeled as being somewhat random. Multipath is usually much more of a problem if a direct LOS path does not exist between the transmitter and the receiver. In this scenario, the change in the magnitude of the received vector as compared to the mean value of the magnitude of the received vector is small, resulting in a Ricean distribution (Figure 1.6). Figure 1.6b shows the vector space representation of the multipath reception in the presence of a LOS path. The vector represents the resultant vector from the LOS path (1) and the multipath receptions (2), (3), and (4). The magnitude of vector represents the mean value of the possible resultant vectors. The area of the circle indicates the 50% contour for this Ricean distribution.[4] It is clear from this figure that a multipath response may not affect the decision variable significantly in such a scenario.

---

[4]Ricean and Rayleigh fading models are the most common fading models applied to analyze propagation in indoor environments. The names of these fading models are derived from their underlying probability distribution function (PDF) statistics. A Rayleigh fading typically occurs if there are several indirect propagation paths between the transmitter and the re-

Figure 1.7a displays the multipath channel in the absence of an LOS path. Figure 1.7b shows the vector space representation of such a response. Vectors (2), (3), and (4) represent the reflected signals at the receiver. Vector (1) represents the intended LOS signal which has been interrupted and reflected multiple times by the scatterers. Vectors (1'), (2'), (3'), and (4') represent the vectors used to find the resultant vector, **a**. It is clear that vector **a** is very small in magnitude, resulting in a high probability of error at the slicer. For large number of scatterers, the channel can be modeled to have a Rayleigh distribution, with about 10% probability of a resultant vector with a magnitude less than half the magnitude of the mean. Note that in this case the mean ±25% contour in the vector space is not a circle because of the asymmetry of the Rayleigh density function about its mean value.

In a typical indoor environment (office, home, etc.) root-mean-square (RMS) delay spreads[5] of 50 to 75 ns can be observed. The worst case RMS delay spreads in these environments can be as large as 150 ns. In order to establish a traditional high data rate communication in such an environment, a very high symbol rate corresponding to a short symbol duration would be required. The larger the value of the RMS delay spread as compared to the symbol duration, the more intersymbol interference (ISI) would be generated. ISI can be corrected in the digital domain, but very high speed and typically high power consumption time-domain equalizers would be needed.

With an understanding of multipath, the benefits of OFDM coding can now be discussed in more detail. OFDM coding is a technique that can be quite powerful in reducing the effects of multipath on high speed communications.

With OFDM, the transmitted data are modulated onto multiple subcarriers. This is accomplished by modulating the subcarriers' phase and amplitude. As such, the original high data rate stream is split into multiple lower rate streams and then mapped on to the available subcarriers (which are multiples of a given frequency) and then combined together using an in-

---

ceiver with none of the paths being a dominant path (i.e., with distinctively larger magnitude than the others). In this situation, the received signal is comprised of the sum of multiple independent random variables and at the limit can be approximated as having a Gaussian distribution function. In reality, Rayleigh fading is really a worst case in which no path dominates. However, since Gaussian PDFs are very well understood and can easily be modeled mathematically, they present a convenient mathematical tool for analyzing the worst case propagation characteristics. On the other hand, Rayleigh fading typically applies if a dominant propagation path (such as a LOS path) between the transmitter and the receiver exists. In this case the PDF is "centered" around the magnitude set by the dominant propagation path.

[5]RMS delay spread is defined from the characteristics of the delay spectrum of a stochastic process. It can be thought of as an indication of the delay between the earliest arriving "rays" and the latest arriving rays.

verse fast Fourier transform (FFT) operation. In creating N parallel transmit streams, the bandwidth of each stream is reduced by a factor N that can be selected in such a way that the RMS delay spread of the channel is much less than the symbol period. This results in a significant reduction in the ISI. A well-designed OFDM system does not therefore require a time-domain equalizer.

The transformations utilized by OFDM are the discrete Fourier transform (DFT) and the inverse discrete Fourier transform (IDFT). The orthogonality of the OFDM signal is obtained through the use of multiples of the subcarrier frequency over an integer cycle which is an inherent property of the DFT and IDFT transformations.

In Figure 1.8 a single subcarrier is displayed in the frequency domain. The OFDM signal is constructed by the summation of multiples of such single subcarriers, as shown in Figure 1.8 (this is an example with five subcarriers). It is clear from this figure that the subcarriers are allowed to have overlap not only with their adjacent subcarrier but also virtually with all of the other subcarriers.

For those familiar with the CDMA technique utilized in many of today's cellular phones, the following analogy may be useful. The construction of an OFDM signal with multiple sinusoidal subcarriers is somewhat similar to the construction of a CDMA signal using orthogonal Walsh codes (Walsh codes are a family of orthogonal codes which are based on "square waves" rather than sinusoids). The main difference between CDMA and OFDM is that in the case of CDMA the orthogonal Walsh codes are primarily used as a means for multiple access,[6] whereas the orthogonal sinusoids in the OFDM coding are primarily used to gain immunity to multipath.

The fact that subcarrier overlaps are allowed enables the spectral efficiency of an OFDM-coded signal to be increased. It is easy to see that with no subcarrier overlap the same number of subcarriers (which is related to the amount of data being communicated) would occupy a much wider spectrum. This would clearly reduce the spectral efficiency. This concept is shown graphically in Figure 1.9 in a simplified diagram.

The obvious question that may arise is the potential interference caused by the overlapping of the subcarriers. However, due to the inherent orthogonality of the subcarriers of the OFDM signal, the peak of each subcarrier occurs at the null of all other subcarriers, as seen in Figure 1.8. Under ideal conditions, this would mean that the subcarriers do not interfere with one another.

Unfortunately, under real-world conditions, various impairments could cause the perfect orthogonality of the subcarriers to be violated. These in-

[6]Note that CDMA also provides immunity to multipath due to the spreading of the signal.

Figure 1.8 Construction of OFDM signal from its individual components (subcarriers). Note the tight "packing" of the subcarriers and the spectral efficiency achieved. Also note that each subcarrier's peak occurs when the other subcarriers are at a null.

clude impairments such as phase noise, quadrature imbalances, distortion, and uncorrected frequency offsets. The location of each subcarrier's peak would shift relative to the other subcarriers in such a way that the peak of one subcarrier would no longer be aligned with the null of the other subcarriers. Such impairments would give rise to "intersubcarrier interference." These impairments and their impact on the OFDM signal and the overall system will be studied in great detail in Chapter 3.

As any good engineer would guess, an OFDM-coded signal could not have all these great properties without some trade-offs.

Probably the biggest "difficulty" with using OFDM-coded data is that it tends to generate very large peak-to-average ratio (PAR) signals. The large

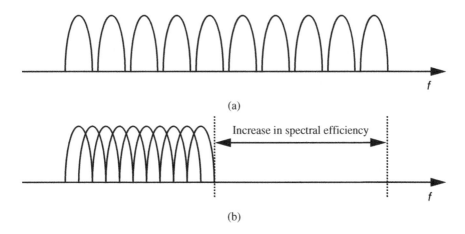

Figure 1.9 Increasing the spectral efficiency of the modulation by using the orthogonal properties of the OFDM signal and packing the subcarriers and their associated data content closer to one another.

PARs significantly complicate the design of the radio and the mixed-signal blocks. The signal path will have to be designed with much more severe linearity constraints than traditional non-OFDM modulations. In particular, on the transmit signal path, the design of the power amplifier becomes quite challenging. Not only is designing high linearity power amplifiers (required by OFDM modulation) quite challenging, but such amplifiers have much worse efficiencies than their nonlinear counterparts.

The topic of the high PAR OFDM-modulated signal and its implications on the power amplifier design will be covered in more detail in Chapter 3.

Now that the general concept of OFDM has been introduced, some of the specifics of 802.11a/g OFDM coding will be discussed.

The 802.11a/g OFDM signal is constructed from 52 total subcarriers, as shown in Figure 1.10. These subcarriers are indexed from −26 to +26, with the zeroth subcarrier eliminated. Out of the 52 subcarriers, 48 are dedicated to carrying the desired data (payload), and 4 of the subcarriers are designated with the task of carrying the "pilot" information.

The subcarrier index numbers for the pilots are −21, −7, 7, and 21. The pilot subcarriers are always modulated in binary phase shift keying (BPSK)[7] format, which is a very simple but robust modulation. The pilot tones are primarily used to help establish a robust "link" before the reception of the desired data (payload) can begin. As such they allow the receiver to set the proper gain, track and correct the carrier frequency offsets, adjust and correct the analog-to-digital conversion (ADC) sampling frequency offsets, and so on. If these tasks are not done properly, the entire packet is likely to be lost, and the effective throughput of the link is significantly reduced. The BPSK modulation, due to its inherent simplicity, is quite robust to various analog and channel impairments such as multipath distortion, phase noise, and quadrature imbalances. This is the reason for transmitting the pilot subcarriers in BPSK format.

The 802.11a/g OFDM subcarriers are spaced 312.5 kHz apart and occupy an overall channel bandwidth of 16.25 MHz,[8] which occupies a baseband bandwidth of −8.125 to +8.125 MHz. The zeroth subcarrier has been eliminated in the 802.11a/g standard and is not used as a pilot or payload subcarrier. This fact has very important implications in the choice and design of the radio architectures used for 802.11a/g solutions. This topic will be discussed in detail later in the book.

The channel-to-channel spacing in the 802.11a standard is 20 MHz. In the 802.11g standard this spacing is set to 25 MHz. The difference between

---

[7]BPSK is the simplest form of the phase shift keying (PSK) modulation family. It is also the same as the simplest form of a quadrature amplitude modulation or QAM-2.

[8]52 subcarriers × 312.5 kHz/subcarrier = 16.25 MHz.

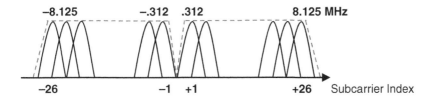

Figure 1.10 Construction of IEEE 802.11a/g OFDM signal from 48 data and 4 pilot subcarriers.

the occupied modulation bandwidth (16.25 MHz) and the channel-to-channel spacing is used to reduce the effects of adjacent channel interference which occur due to imperfections in the transmitter and the receiver.

1.8   802.11a/g: DATA RATES

The various data rates allowed in the 802.11a/g OFDM mode are shown in Table 1.3. As can be seen, the data rates range from 6 to 54 Mbps. The data rates are varied from the highest to the lowest rates by changing one or both of the following modulation-related parameters: (a) modulation order and (b) coding rate.

The modulation order is the primary tool used to adjust the data rate for 802.11a/g. At the higher order modulations, for a given transmit power and with everything else being the same, the spacing between the neighboring constellation points on a constellation diagram is less than those of lower order modulations. This makes the modulation much more susceptible to im-

Table 1.3   802.11a/g Data Rates, Modulation Types, Coding Rates, and Required Sensitivity Levels Set by Standard

Data Rate (Mbps)	Modulation	Code Rate	Sensitivity Requirement (dBm)	State-of-the-Art Chip Sensitivity (dBm)[a]
6	BPSK	$\frac{1}{2}$	−82	−94
9	BPSK	$\frac{3}{4}$	−81	−92
12	QPSK	$\frac{1}{2}$	−79	−90
18	QPSK	$\frac{3}{4}$	−77	−87
24	QAM-16	$\frac{1}{2}$	−74	−84
36	QAM-16	$\frac{3}{4}$	−70	−82
48	QAM-64	$\frac{2}{3}$	−66	−76
54	QAM-64	$\frac{3}{4}$	−65	−74

Note: Representative state-of-the-art sensitivity levels are also specified.
[a]Using hard Viterbi decoding can improve the sensitivity of higher order modulations by as much as 2.5 dB.

pairments such as circuit noise, phase noise, and in-phase/quadrature phase (I/Q) imbalance.

The code rate determines the amount of redundancy and hence robustness built into the modulation. The closer the coding rate to unity, the less the amount of redundancy built in, and the higher the data rate (the data are not "wasted" for the sake of redundancy).

The coding rate is another tool utilized to adjust the data rate. Typically, however, the change in data rates as a result of a change in the coding rate is much smaller than that of changing the modulation order. This is because coding rates much larger than $\frac{5}{6}$ do not provide enough redundancy to be useful and are therefore not typically used in practice. Several examples of changing the data rate by utilizing various coding rates are shown in Table 1.3.

In a real system, the control of the actual data rate selected by the link is done through the media access controller. The goal of MAC is to establish the fastest (but reliable) link possible. As such, it typically starts at the highest data rate and tries to establish a robust link. If it fails to do so, it will drop the rate to a lower rate and retry. It will continue this process until it establishes a link or determines that no link can be established. Detailed discussions of the MAC layer are beyond the scope of this book and the interested reader can refer to the references.

The IEEE 802.11a/g standards require any system that claims compatibility to the standard to be able to maintain certain minimum sensitivity levels (ranging from –65 to –82 dBm for the various data rates). The minimum required sensitivity level by the standard for the various data rates is listed in Table 1.3. Today's systems can significantly outperform the specifications for sensitivity which have been set by the standard. Table 1.3 also shows examples of the capabilities of today's state-of-the-art integrated solutions referred to the input of the chips. In general, the performance of the state-of-the-art solutions is about 10 dB superior to those required by the standard. It is important to note that several assumptions have been made in specifying the sensitivity of the state-of-the-art solutions: (a) the sensitivity numbers specified are referred to the chip input (i.e., the board losses, which can range from 1 to 3 dB are not accounted for); (b) no "external" (nonintegrated) low noise amplifiers (LNAs) are assumed in front of the receiver chip; and (c) hard Viterbi decoding is assumed for the baseband section of the receiver.[9]

It is interesting to note the inverse relationship between the data rates and the minimum sensitivity of the various modes of operation shown in Table

---

[9]A soft Viterbi decoder would improve the performance numbers specified by as much as 2.5 dB for the higher data rates as compared to the numbers shown in Table 1.3. It will improve the sensitivity for the lower data rates marginally, however, since at the lower data rates the sensitivity is often limited by the problem of "detection" (i.e., whether there is a packet present).

1.3. As the data rates are increased (through increasing modulation order or by using higher coding rates), the minimum sensitivity level suffers. Given the explanation earlier, this should be rather obvious and is related to the larger SNR required by the higher data rates. In other words, as the data rate increases, a higher received power level is required in order to be able to receive the signal (assuming noise levels stay constant). The absolute level of the SNR required for each data rate is dependent on various factors (soft versus hard Viterbi decoding as an example) but is in all cases higher than that of a lower data rate (all else being equal).

Although not shown in Table 1.3, it is a similar situation on the higher end of the power range. The 802.11a/g standards do specify the minimum high end power rate that the receiver should be able to receive (−30 dBm). However, unlike the minimum power level requirements, at the high end the power levels are not specific to each data rate. In reality, though, the higher data rates are much more susceptible to "high power impairments" such as nonlinearities in the receiver (and transmitter). So the receiver would quite likely be able to tolerate much higher receiver power levels for a 6-Mbps link than a 54-Mbps link. This should be obvious by considering the fact that high power impairments such as nonlinearities cause the constellation points on a constellation diagram to deviate from their ideal point and get closer to the neighboring constellation points. Since for a given transmit power the spacing between the constellation points on a high order modulation is larger than that of a low order modulation, the low order modulation would be able to handle much more nonlinearities before it causes an error.

As a side note, given our knowledge of the 802.11a/g and that the duration of each symbol is 4 μs, we should now be able to calculate each one of the data rates listed in Table 1.3. For example, the 54-Mbps data rate can be calculated as follows:

$$48 \text{ (data subcarriers)} \times 6 \text{ (bits/symbol for QAM-64)}$$
$$\times \tfrac{3}{4} \text{ (code rate)} \times \tfrac{1}{4} \text{ (μs)} = 54 \text{ Mbps}$$

For the 6-Mbps data rate

$$48 \text{ (data subcarriers)} \times 1 \text{ (bits/symbol for BPSK)}$$
$$\times \tfrac{1}{2} \text{ (code rate)} \times \tfrac{1}{4} \text{ (μs)} = 6 \text{ Mbps}$$

It is important to make one final point on Table 1.3. For 802.11g, this table only shows the OFDM-related rates. As mentioned earlier, 802.11g is backward compatible with 802.11b and as such is capable of operating at all the lower data rates (11, 5.5, 2, 1 Mbps) at which 802.11b is capable of operating.

1.9   802.11a/g OFDM PACKET CONSTRUCTION

The physical layer convergence protocol (PLCP) layer allows for a common MAC layer to be used for many 802.11 WLAN PHY substandards. The construction of an 802.11a/g OFDM packet is shown in Figure 1.11. As shown in the figure, the packet is comprised of four basic components. The first piece of the symbol is what is known as the "short" preamble. The short preamble is always 8 μs long in duration. During the short preamble, tasks such as automatic gain control and coarse frequency offsets are calculated and adjusted for in the baseband chip. The short preamble is followed by a "long" preamble, which is also 8 μs long.[10] During the long preamble, tasks such as channel estimation, fine frequency offset adjustments, and timing recovery are performed. Again note that the short and long preambles are always communicated in BPSK in order to maintain robustness.

The field that follows the preambles is called the "signal field." It is here that the information relating to the modulation order, coding rate, and packet length is carried. The short and long preambles along with the signal field constitute the PLCP protocol data unit (PPDU).

The actual payload (which carries the desired data to be communicated) follows the signal field. So the PPDU plus the user data together constitute the full packet.

Note that the PLCP preamble is made of 12 OFDM symbols, the signal field is made of 1 OFDM symbol, and the user data are made of a variable number of OFDM symbols.

1.10   802.11 SYSTEM REQUIREMENTS

We will now shift our focus and discuss some of the more important system requirements for the 802.11 standard.

1.10.1   Receiver Sensitivity

We will start by discussing the receiver sensitivity requirements. In general the receiver sensitivity is affected by many factors, including the design of the radio and the baseband PHY layer.[11] Examples of the latter case are the choice of the Viterbi decoder (soft vs. hard), the design of the channel estimator, and the gain control algorithm.

---

[10]Note that both the short and long preambles are 8 μs in duration. However, the short preamble is constructed of 10 smaller (800-ns) segments—hence the word *short.*

[11]Purist radio designers find the latter to be hard to believe!

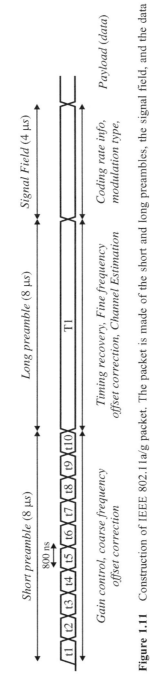

Figure 1.11 Construction of IEEE 802.11a/g packet. The packet is made of the short and long preambles, the signal field, and the data payload.

On the radio side, the factors impacting the minimum sensitivity levels are related to the data rate selected. In general, under low data rate conditions (e.g., 6 Mbps), the sensitivity of the system is primarily set by the receiver noise figure and cochannel interference.[12] For the most part other impairments such as quadrature imbalances would have minimal impact at the sensitivity levels of the lower data rates.

The situation is quite a bit more complex at the higher data rates (such as 54 Mpbs), however. It is clear that noise figure and cochannel interference will still impact the sensitivity levels, but other impairments such as phase noise, quadrature imbalance, transmitter error vector magnitude (EVM), center-frequency inaccuracies, filter corner inaccuracies, multipath, sampling frequency inaccuracies, and gain control inaccuracies will also enter the picture and impact the sensitivity levels. Once again, it is clear that the higher data rates would require a higher SNDR (signal to noise plus distortion level) to be able to operate properly and would therefore have a more limited power range in which they can operate robustly.

The discussions of the previous few paragraphs assume that there are no significant interferers present. Under interference-dominated conditions (conditions in which the desired signal level is significantly smaller than an interfering signal such as a large adjacent channel signal), the linearity of the receiver will also become a factor in determining the sensitivity of the system. Under interference-dominated conditions, a "smart receiver" would need to be able to detect the interference condition (through the use of proper RSSIs, for example) and set the front-end gain control accordingly. This often translates into a trade-off between linearity and noise figure. In general, the gain of the front end would have to be set to the highest level possible (corresponding to the lowest noise figure possible) while avoiding significant nonlinearities in the receiver chain. So, in summary, in interference-dominated conditions, in addition to the linearity of the receiver, all the factors mentioned above for the non-interference-dominated conditions are important and must be considered in the design.

1.10.2   Transmitter Error Vector Magnitude

Another system requirement specified by the 802.11a/g standard is the transmitter EVM, which, in general is a single scalar number that is an indication of the modulation quality of the signal. To calculate the EVM, one needs to compare the actual symbols with their ideal impairment-free sym-

---

[12]Cochannel interference is caused by users in adjacent cells operating at the same frequency as the user. Since the user data in the adjacent cell is uncorrelated to the user data in the current cell, this would cause a noiselike effect on the user and would degrade the sensitivity levels.

bols on the constellation diagram and compute the error vectors as shown in Figure 1.12. The real symbol will have a different phase and amplitude as compared to the ideal symbol constellation points. Systematic and deterministic errors would simply offset the real constellation points as compared to the ideal ones. Nonsystematic impairments such as noise, however, would cause an "error ball" or "error cloud" of uncertainty in the constellation points about the ideal constellation points.

Mathematically, for a given symbol, EVM is defined as

$$\mathrm{EVM} = \sqrt{\frac{\sum\limits_{i=1}^{M} |Z(i) - R(i)|^2}{\sum\limits_{i=1}^{M} |R(i)|^2}}$$

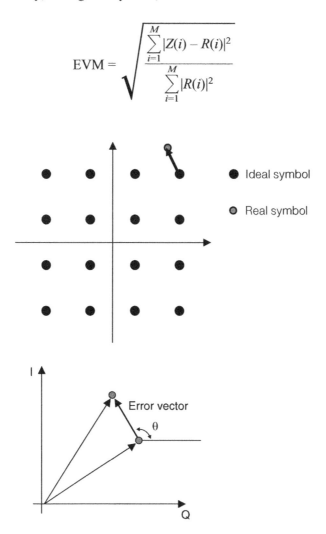

Figure 1.12   Pictorial description of the concept of EVM for 16-QAM constellation diagram. The detail (zoom in) of the EVM calculation applied to a single constellation position is also displayed.

where Z is the measured signal, R is the reference (ideal) signal, M is the number of measurements, and i is the measurement index. Various standards allow for some form of tracking of the symbols (such as timing recovery and frequency offset corrections) before applying this equation to measure EVM. This definition can be extended to all the symbols of a modulation by averaging over all these symbols.

It can be seen from the definition that the EVM is an estimate of the magnitude of the error signal as compared to the magnitude of the ideal signal. As such, it is clear that that the maximum value for the EVM is 1, or 100%, and that the minimum value of the EVM is 0. It is clear that the smaller the value of the EVM in percent, the higher the quality of the modulated signal. The EVM can be expressed in decibels as 20 log(EVM). For example, 1% EVM can be expressed as –40 dB EVM.

However, in some cases different standards specify the EVM definition in slightly different terms. In the case of 802.11a/g OFDM rates the EVM is defined as the RMS of the error vectors over all symbols. In contrast, in the case of 802.11b/g CCK rates, the EVM is calculated using the peak values.

The EVM is affected by a variety of impairments. These include nonlinearities, phase noise, quadrature imbalances, and filter shapes and bandwidth. The reason that the EVM is so commonly used as a measure of the quality of the transmitter is that it is essentially impacted by many such impairments. Of course, one should note that there are impairments that may not impact the EVM as measured by a vector signal analyzer (VSA), and conversely there are impairments that may impact the EVM as measured by a VSA but may not have a significant impact on the packet error rate as measured by the actual receiver. More on this point will be discussed later.

For the 802.11a and g standards, at 54 Mbps, the transmitter EVM (or transmitter quality) is required to be a minimum of –25 dB. On the other hand, there are no specifications given for a receiver EVM. The receiver is completely specified by the specification of the sensitivity for a given data rate. If there are impairments that would affect the receiver EVM, they will quite likely be evident in the sensitivity measurements of the receiver, especially at the higher data rates, and this will impact only that receiver. On the other hand, a problem in the quality of the transmitted signal can cause another user (with potentially a perfectly good receiver) to be unable to operate at the proper data rate. This is the reason that the standard enforces the EVM on the transmitter but not on the receiver.

1.10.3 Transmitter Spectral Mask

Another 802.11 system requirement on the transmitter side is the satisfactory passing of the spectral mask. To facilitate the discussion of spectral mask, four observations need to be made here:

1. A linear system is one that faithfully replicates the input signal by a constant multiplier (gain or attenuation) without the introduction of any frequencies (harmonics, intermodulations, etc.) that did not exist in the incoming signal. Of course, a constant value can be added to the output. The relation $V_{out}(t) = kV_{in}(t) + c$ describes the transfer function of a linear system, where k and c are constants. If a sinusoid of frequency f_1 is applied as $V_{in}(t)$, the output spectrum will only have components at f_1 and possibly at the direct current (DC). No other frequency components will be present.

2. A nonlinear system, in contrast, is capable of generating frequencies in the output spectrum that did not exist in the incoming signal. A nonlinear system may, for example, be represented by the following transfer function: $V_{out}(t) = kV_{in}^2(t) + c$. If a sinusoid of frequency f_1 is applied as $V_{in}(t)$, the output spectrum will have components at f_1 and $2f_1$ and possibly at DC (this can be seen by applying basic trigonometric identities). In this example, if two sinusoids are applied at the input, one with frequency f_1 and another with frequency f_2, tones at $2f_1$, $2f_2$, $f_1 - f_2$, and $f_1 + f_2$ will be present at the output. It is clear that frequencies at the output which did not exist at the input have been generated.

3. A constant-envelope modulation is a modulation that has no peaks and valleys observed in the transient waveform of the modulated signal. In other words, the envelope of the carrier does not change with a change in the modulated signal. This class of modulations includes modulations such as FM (frequency modulation) and FSK (frequency shift keying).

4. A nonconstant-envelope modulation is one that has peaks and valleys in the transient waveform of the modulated signal. In this class of modulations, the amplitude of the envelope of the modulated signal can vary as a function of time. This class of modulations includes modulations such as AM (amplitude modulation) and QAM (quadrature amplitude modulation).

In general, all real systems are at least weakly nonlinear (i.e., perfectly linear systems do not exist in the real world). Also, in general, constant-envelope modulations are significantly less spectrally efficient than their nonconstant-envelope counterparts. On the other hand, as will become apparent by the argument below, nonconstant-envelope modulations are much more forgiving of nonlinearities in the system.

All forms of modulations used by the 802.11 standard possess a relatively high spectral efficiency and have a high PAR.

Now that we have some of the basics out of the way, we can discuss the concept of spectral regrowth in more detail. When a modulated signal of bandwidth W (constant envelope or not) is passed through a system with no nonlinearities, a signal with the same W bandwidth is obtained at the output. For example, assuming a "brick-wall"-shaped signal was applied at the input of such a system, the signal at the output would maintain the same brick-wall shape and bandwidth (and would of course have no harmonic outputs).

If the modulated signal is of the constant-envelope type and it is passed through even a nonlinear system, it maintains its bandwidth.[13]

In reality, however, nonlinearities are ever so present. Therefore when a non-constant-envelope signal (especially one with large PARs) is passed through such a system, even- and odd-order harmonics of the input signal are generated, creating harmonic distortion (HD) at the output. These components are typically out of band and can be filtered out. However, if there is more than a single CW tone present at the input, intermodulation (IM) distortion terms will also be generated. Specifically, the odd-ordered intermodulation terms (IM3, IM5, IM7, etc.) are the ones that result in spectral regrowth.

The concept of intermodulation distortion (IMD) will be discussed in more detail in Chapter 3, but for now it suffices to note that IM terms can generate spectral components that fall very close to the frequencies of the input signal. For example, for a two-CW tone input with frequencies f_1 and f_2, the problematic third-order IM products (IM3) fall at $2f_1 - f_2$ and $2f_2 - f_1$. Since f_1 and f_2 are presumably very close in frequency, the $2f_1 - f_2$ and $2f_2 - f_1$ terms will also fall fairly close to the frequency range and will be difficult or even impossible to filter out. It is easy to imagine with an input signal with many CW input tones how the intermodulation terms would generate a spectral intermodulation floor within the band of interest and its vicinity.[14] A modulated signal with a nonconstant envelope for the purposes of this discussion can be considered as a multitone CW signal with frequency components across the modulated bandwidth.

Although these IMD terms are present *within* the bandwidth of the modulated signal, their amplitudes are significantly smaller than the amplitude of the desired spectral component of the modulated signal and can therefore typically not be observed by looking at a simple power spectral density plot

---

[13]Note that if a modulated signal with abrupt phase transitions that has constant envelope is passed through a band-limiting operation (e.g., filtering), it will no longer have a constant envelope and will therefore require somewhat linear amplification in order to avoid the generation of spectral regrowth.

[14]This intermodulation floor is sometimes referred to as the spectrum "grass" or FFT grass. This is because, in looking at the FFT results of such a multitone simulation, the FFT bins of the area immediately outside the bandwidth of interest will show FFT components that have grown above their normal levels.

or a spectrum analyzer output (they do certainly take a hit on EVM, however). These intermodulation components, however, will result in the creation of undesirable spectral components (as observed on a spectrum analyzer or a power spectral density plot) in the immediate vicinity of the modulated signal. These undesirable spectral components generated by the passing of a nonconstant-envelope modulated signal through a nonlinear system is commonly referred to as spectral regrowth.

With the description provided in the previous paragraphs it is easy to understand why large PAR signals are more susceptible to IMD and hence can create large spectral regrowth components as a result of passing through a nonlinear system.

To summarize, when a modulated signal is passed through a nonlinear system, its bandwidth is broadened by odd-order nonlinearities. This is caused by the creation of mixing products between the individual frequency components of the modulated signal spectrum.

In a typical system, spectral regrowth is typically of concern in the transmitter side. Within the transmitter, spectral regrowth is typically caused by the most nonlinear component of the transmit chain. This component is almost always the power amplifier.

Spectral regrowth can cause several problems in a system. In a full-duplex system, for example, a client's transmitter can generate enough out-of-band power due to spectral regrowth to saturate the client's own receiver (note that a transmitter in a typical full-duplex system can be transmitting power levels that are orders of magnitude larger than what the receiver is trying to receive, and therefore it is fairly easy to saturate this receiver). Of course, WLAN systems, in particular 802.11, are not based on a half-duplex operation, and therefore this particular potential impairment caused by spectral regrowth is not an issue for a WLAN system.

A similar issue can be observed when spectral regrowth of one transmitter causes the desensitization of the receiver of a different system in the same communication appliance. For example, the transmitter of a WLAN system operating at 2.4 GHz can desensitize the receiver of a wireless CDMA cellular receiver in a cell phone hand set.

Another problem caused by spectral regrowth is to create interference with adjacent channels. Typically, the spectrum is a very precious commodity. The channels are therefore typically packed together in order to maximize the efficiency of the spectrum usage. Excessive spectral regrowth can therefore cause interference with the adjacent channels (this is the primary concern with spectral regrowth in stand-alone 802.11 applications such as computer laptops). This is the primary reason why the standards and the FCC require conformance to a spectral mask.

Take the case of the 802.11a spectral mask requirements as an example. As discussed earlier, the 802.11a modulated signal is comprised of 52 total subcarriers with modulated data around each of these subcarriers. Also, as stated earlier, such a signal possesses a very high PAR. As this signal is passed through a nonlinear system, these subcarriers can nonlinearly interact with other subcarriers, creating intermodulation components. The data content of each subcarrier can even nonlinearly interact with itself, causing IMD. All of this can cause spectral regrowth. So an ideal brick-wall filtered OFDM-modulated signal (with a bandwidth of 16.25 MHz) could look like what is shown in Figure 1.13 after it goes through a real transmitter with a much wider bandwidth. As shown in Figure 1.13, the 802.11a spectral mask requires that the spectrum of the transmitted signal be more than 20, 28, and 40 dBc below the peak of the modulated signal at offset frequencies of 11, 20, and 30 MHz, respectively, away from the center of the band.

Finally, it is important to note, that the spectral regrowth due to the nonlinearities (of typically the power amplifier) *should* be the limiting factor in achieving the (close-in) spectral mask requirements of a well-designed system which has a fairly high transmit power requirements (such as the 802.11). However, many other factors can cause spectral mask violations. These include insufficient baseband analog or digital filtering of the digitally generated modulated signal, the quantization noise levels of the digital-to-analog converters, and the phase noise of the RF phase-locked loop (PLL).

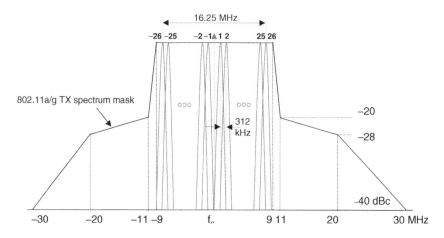

Figure 1.13 IEEE 802.11a/g channel construction from OFDM subcarriers and required transmitter spectral mask.

The subject of spurious emissions is also closely related to the issue of spectral masks. For example, the 802.11 standard specifies that the transmitter local oscillator (LO) feedthrough[15] should be limited to less than 15 dB below the level of the desired transmitted signal power for the 802.11a and 802.11g standards. For the 802.11b standard, the level of LO feedthrough needs to be at least 25dB below the desired signal level. It is important to note that quite often LO feedthrough can exceed the standard specified limits without affecting the transmitter error vector magnitude (see next section). However, excessive LO feedthrough can cause problems on the receiver side, especially for direct-conversion receivers (by generating large-baseband DC offsets).

Another type of spurious emissions is synthesizer-generated and LO-generator-type spurs. For example, the synthesizer can have an excessively large reference feedthrough which can easily violate the spectral mask of the system and/or cause interference with other users that are operating at the adjacent channels. Another example would be the case of the first or second LO of a superheterodyne transmitter leaking to the output and violating an FCC mask for a restricted band. As a final example, quite often in direct conversion systems, the voltage-controlled oscillator (VCO) frequency is selected to be different than the output frequency of the transmitter in order to reduce the pulling effects on the VCO. In such cases the VCO frequency can leak to the output and potentially violate the FCC limit in a restricted band.

The FCC requirements can be divided into two main categories: conductive requirements and radiative requirements. Conductive measurements are performed by connecting a cable to the antenna port of the device under test and measuring the power spectral density of the transmitted signal at various frequencies out of the 802.11 bands. On the other hand, radiative requirements are conducted by measuring the received power at various frequencies at certain defined distances and with specified antennas. The interested reader can refer to FCC documents for more details.

1.11 VECTOR SIGNAL ANALYSIS

A very powerful tool in analyzing a digitally modulated signal is a VSA. A VSA is even more useful in analyzing digitally modulated signals that are

---

[15]The subject of LO feedthrough will be discussed in more detail in Chapter 4. For now, it is sufficient to note that LO feedthrough is the continuous wave (CW) tone observed in the middle of the transmitted band as observed at the RF output. The two most common sources of LO feedthrough are DC offsets in the baseband analog circuitry of the transmitter and, in the case of direct-conversion transmitters, direct LO coupling to the output of the transmitter.

further mapped by OFDM. This is because in many cases the impact of impairments on an OFDM-modulated signal would be quite different (and often more complicated) than a similar impairment's impact on a "single-carrier"-based (i.e., non-OFDM) modulated signal. Further, typically the impact of various impairments on single-carrier-based modulated signals is more intuitively clear. We will elaborate on this issue in much more detail in Chapter 4.

Vector signal analysis is a tool to relate analog impairments to system requirements. In its simplest form one would look at the constellation diagram of the digitally modulated signal as shown in Figure 1.14. This figure shows a very high quality 64-QAM (an array of 8×8 blue constellation points) 802.11a signal, with an EVM of approximately –40 dB.[16] The high quality of the modulation is apparent in the tightly packed dots at the intersection of the circles. As described earlier in discussing the concept of the EVM, these tightly packed dots are an indication that the actual received symbols are quite close to their ideal values. As the signal quality degrades, the dots get larger and "fuzzier" and turn into "balls," as seen in Figure 1.15. At the extreme the various constellation balls start intruding on the adjacent neighbor constellation balls, causing packet errors and degrading the link quality. Note that the black dots in Figure 1.14 in the center left and the center right are associated with the 802.11a pilot tones, which are always transmitted in BPSK format.[17] In particular, these constellation points are quite useful in identifying certain kinds of impairments in the system. More details on this topic will follow later in this book.

So, by looking at a constellation diagram, one can recognize a "good"-quality signal such as that shown in Figure 1.14. But what if the constellation diagram shows a signal with a relatively poor quality (e.g., that of Fig. 1.15)? How would the designer go about finding the problem and remedying it? How would one determine if the problem is due to spurs, phase noise, quadrature imbalance, and so on? This is where, especially in the case of an OFDM-mapped signal, further signal analysis tools and diagrams would be useful.

Figure 1.16 shows an example of an 802.11a-modulated signal (the same signal of the constellation diagram of Fig. 1.14) viewed on a VSA.

---

[16]Note that the 802.11a/g standard requires an EVM of only –25 dB for a a/g 64-QAM transmitted signal; therefore the constellation shown here is 15 dB better than what the standard requires.

[17]The pilot tones are always transmitted in BPSK format. The pilot tones are used to establish the initial link (establish frequency offsets, set packet gain levels, etc.). It is therefore imperative to obtain the highest amount of immunity to impairments present in the system and in the channel and establish a robust link for the payload to be properly decoded.

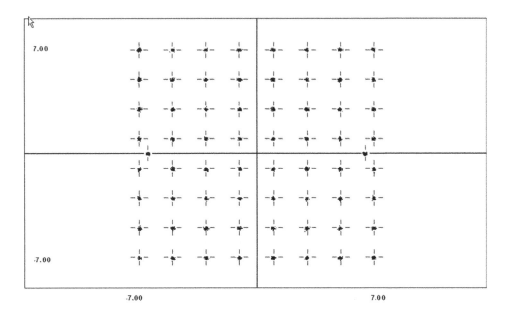

Figure 1.14 Constellation diagram for 802.11a signal with very low (good) EVM (~ –45 dB). This constellation diagram is obtained by feeding the output of a laboratory-class transmitter to a VSA.

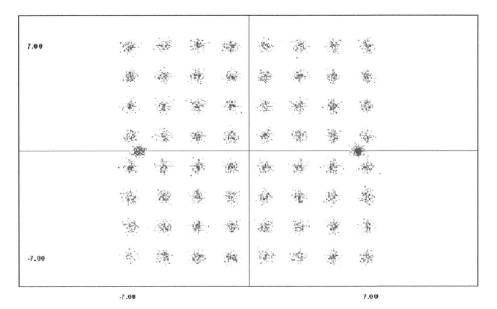

Figure 1.15 Constellation diagram for 802.11a signal when quality of signal is marginally acceptable for 54-Mbps transmission. The EVM of this constellation is ~ –27 dB. Note the "fuzzy" balls that have replaced the well-defined constellation points of Figure 1.14.

In this case, however, the VSA is set up to display the EVM in decibels versus the subcarrier index.[18] In other words, this is an alternative way to view the same signal. As will be shown, this method of observing the signal will shed more insight into certain impairments that may be affecting the system.

The key insight is that a quick glance at Figure 1.16 can provide significant amount of information about the existence (or lack of) analog impairments in the system. Figure 1.16 shows a signal with excellent quality (very low EVM) across all of the subcarriers. Some examples of signals that are impaired in various ways are given in the following figures. The spectrum flatness and group delay associated with this near ideal signal is shown in Figure 1.17.

Figures 1.18a and 1.19b show a signal which has a significant CW spur present at subcarrier 13 (frequency offset of +4 MHz). As can be seen, the EVM for this subcarrier is significantly degraded as compared to other subcarriers. Also note that, by looking at the constellation diagram alone, it would not be possible to pinpoint the reason for the degraded performance. On the other hand, by looking at the plot of the EVM versus the subcarrier, it is quite clear that a large narrowband interference is the source of the degraded EVM. Many sources can contribute to large spur levels. These include harmonics of the crystal oscillator, reference spurs of the PLL, and harmonics of the master clock frequency used in the digital domain of the chip.

Figure 1.20 shows a signal which has fairly significant impairments on the lower index subcarriers.[19] This situation, for example, can arise from excessive flicker noise in the baseband circuitry or as a result of cutting off the low frequency subcarriers by a high pass filter with too high of a corner frequency. The EVM hit can come from the magnitude attenuation due to filtering at the low index subcarriers and/or the group delay variations due to the pole(s) associated with the high pass filter. This situation is not uncommon on 802.11a direct-conversion receivers that use some form of high pass filtering to reject the DC offsets.

Figure 1.21 shows a signal which has significant impairments on the high order subcarriers. This situation is the dual of that described in the previous paragraph but is caused by low pass filters with the poles placed

---

[18]Recall that the 802.11a signal is constructed of 52 subcarriers ranging from index –26 to +26 with subcarrier index 0 eliminated. The subcarriers are spaced 312.5 KHz apart.

[19]Note that the EVM floor in this case is different that the previous examples due to the fact that a different device under test (DUT) was used for this measurement.

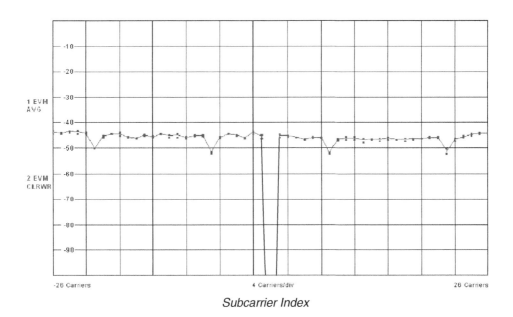

Figure 1.16   EVM versus subcarrier index for constellation plot of Figure 1.14 obtained on a VSA.

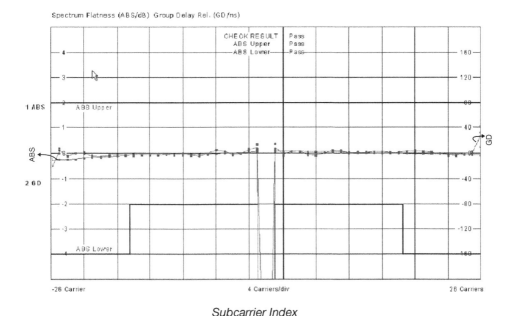

Figure 1.17   Spectrum flatness (ABS, dB) on the left y-axis and group delay variation (GD, ns) on the right y-axis of signal of Figure 1.14.

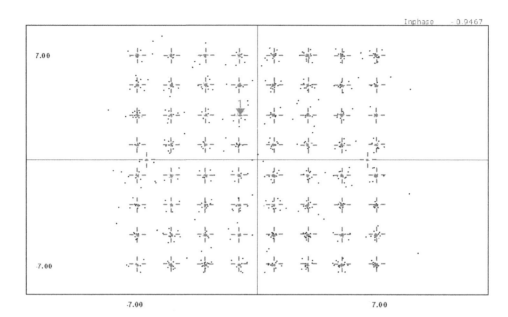

Figure 1.18 Constellation diagram of otherwise excellent quality 802.11a signal with large CW spur. In this case the burst power (of the OFDM signal) is –5 dBm and is centered at 5.24 GHz, and the CW spur has a power level of –35 dBm and is at 5.244 GHz. Measured EVM is –31.6 dB. Notice the out-of-place constellation points on the constellation diagram.

too low. Again, the EVM hit can be due to the magnitude attenuation of the higher order subcarriers (and associated SNR degradation) or due to excessive group delay variations associated with the poles of the high pass filter being placed too low in frequency. A similar situation can arise if a constant group delay variation exists between the I and the Q channels of the received signal. This results in a subcarrier-dependent quadrature imbalance which, if uncorrected, would result in worse EVM at higher subcarriers. The plot of Figure 1.21 represents such a case where a significant delay difference exists between the I and the Q channels. As explained further in Chapter 3, this condition may arise due to mismatches in the analog baseband sections of a receiver. A plot of the amplitude variation of this signal over the subcarrier as well as the group delay variation is shown in Figure 1.22. The constellation diagram for this signal is shown in Figure 1.23. It is again clear that, by looking at the constellation diagram alone, it would be quite difficult, if not impossible, to determine the type of impairment impacting the system.

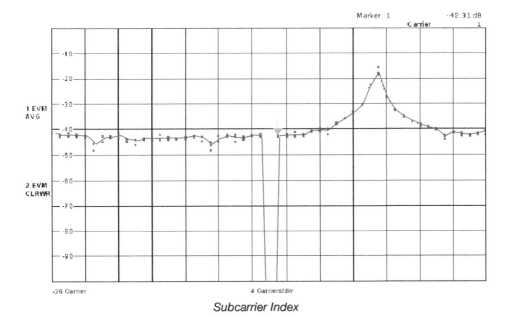

Subcarrier Index

Figure 1.19 EVM versus subcarrier for constellation diagram of Figure 1.18. Note the large degradation in the EVM level at the frequency of the spur. Also notice the "leakage" of the EVM degradation effect on the adjacent subcarriers.

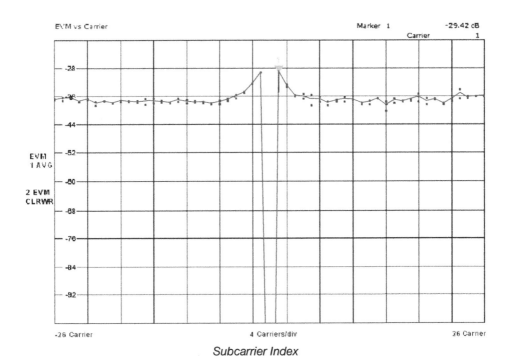

Subcarrier Index

Figure 1.20 EVM versus subcarrier plot of 802.11a signal that shows fairly significant impairments on lower index subcarriers.

39

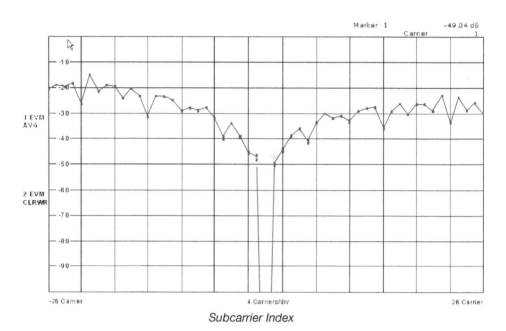

Figure 1.21 Plot of EVM versus subcarrier index for 802.11a signal subject to significant group delay mismatch between I and Q channels.

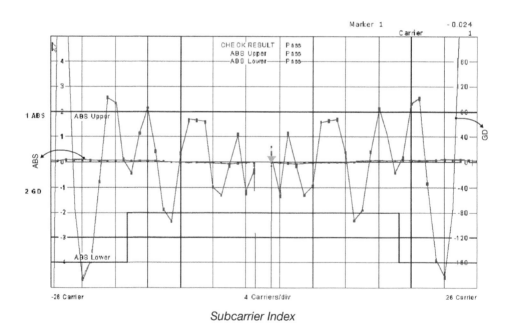

Figure 1.22 Plot of spectrum flatness (ABS, dB, left y-axis), and group delay (GD, ns, right y-axis), versus subcarrier index for 802.11a signal of Figure 1.21,

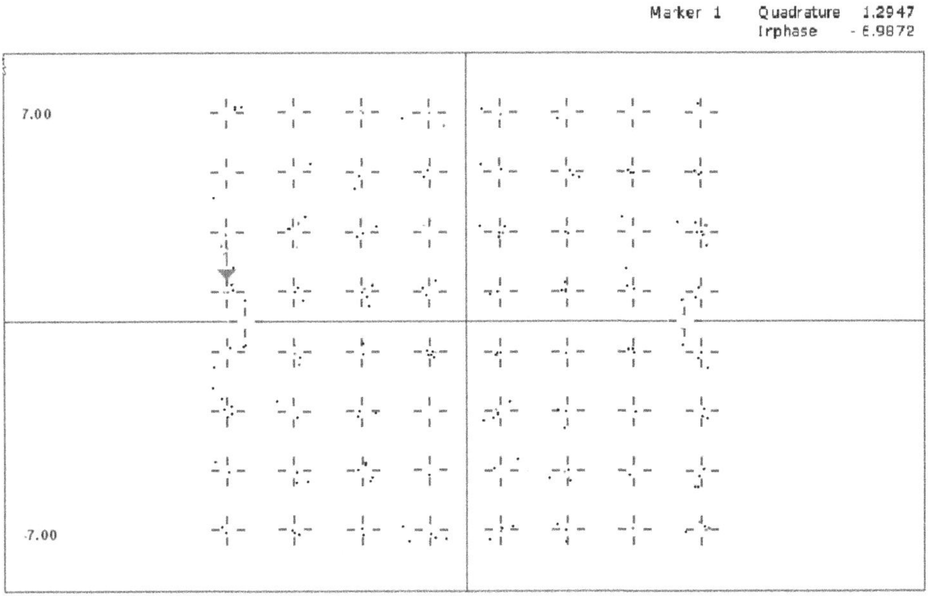

Figure 1.23   Constellation diagram of signal of Figure 1.21.

The examples mentioned above should provide a good perspective on the capabilities of a VSA. Often, the problems could be even further analyzed to very accurately pinpoint the source of the problem. At the very least, this kind of analysis provides the proper hints that the engineers can then use to debug their system. Several more examples of various impairments and methods of debugging them using a VSA will be presented throughout the book.

CHAPTER 2

Radio Receiver and Transmitter Architectures

2.1 ARCHITECTURES

The choice of radio architecture has a significant impact on many aspects of the operation of the system, and it is one of the most important decisions that the system architect has to make. In general, the decision is dependent on many factors, as will be described in some detail in the following pages. These factors include the modulation bandwidth, the existence of information at the center of the modulation, the process technology available, the capability and power consumption of analog-to-digital and digital-to-analog converters (ADCs, DACs) available to the system architect, the risk tolerance of the project, the degree of integration required, the area required, the power consumption required, the performance required, the degree of local oscillator feedthrough (LOFT) acceptable, the amount of LO reradiation acceptable, the quadrature accuracy required, the magnitude of a potential signal at the image frequency, the magnitude of interferers, and, of course, the comfort and experience of the designers with the various architectures. Of course, some of the factors mentioned are more relevant to the receiver architecture, some are more important on the transmitter architecture, and some impact both the receiver and transmitter architecture.

In general, there are three types of common receive and transmit architectures available to the system architect: direct conversion [also known as zero intermediate frequency (IF) or homodyne], low IF, and superheterodyne. Each one of these architectures has its own inherent strengths and weaknesses, and many of the potential issues with each architecture can be overcome with clever architecture or circuit design techniques. Some of these techniques will be discussed in some detail in the following pages.

In the following sections, we will first investigate the receiver architectures and then look at the transmitter architectures. For each receiver and transmitter we will first examine the superheterodyne architecture, which is the architecture that has been around for the longest time and even today the

Wireless LAN Radios: System Definition to Transistor Design. By Arya Behzad
Copyright © 2008 the Institute of Electrical and Electronics Engineers, Inc.

most common architecture in use (if all radios produced today are included; not just the WLAN radios). Then we will discuss the low IF architecture, which is a common choice for certain applications (typically those that have a fairly narrow modulation bandwidth). Finally we will discuss the direct-conversion architecture, which is the architecture of choice for many applications. It is important to note that within these categories some variants exist; however, most, but not all, radio architectures can be generally categorized as one of these three architectures.

2.1.1 Superheterodyne Receiver

Figure 2.1 displays a general superheterodyne receiver architecture with an example frequency planning applied to the 802.11a standard.[20] The signal flow is as follows: The signal is received at the antenna and then passed to a band-select ceramic or surface acoustic wave (SAW) filter. This filter has the task of rejecting any large out-of-band interferers as well as any image signal in order to avoid the desensitization of the RF front end and in particular the LNA. The signal is then amplified by the LNA and sent to a second filter whose primary task is the rejection of the signal at the image frequency as well as any noise present at the image frequency. The image-rejected and band-selected signal is then passed to the RF mixer, which down converts the entire RF band to an IF (in this case 1280 MHz). The RF PLL frequency is selected such that the desired channel falls at the predefined IF frequency (1280 MHz in this case). Now that the signal is down converted to a fixed IF and properly centered, a fixed channel-select filter (in this case a SAW filter[21]) selects the desired channel (or possibly a few channels) while rejecting all the other channels in the band. This significantly reduces the linearity requirements of the blocks that follow the SAW filter.

The level of the signal is then adjusted through the IF programmable gain amplifiers (PGAs) and passed on to the second set of down-converting mixers. Often at this stage the signal is split into the in-phase and quadrature-phase components by using an LO that has quadrature outputs (sine and cosine). The LO frequency is often chosen (as in this example) to take the IF signal directly to baseband (centered around 0 Hz).[22] Any further filtering that may be required is implemented at the baseband frequency at this point, and further gain control is performed as necessary to set the signal level to

---

[20]Blocks associated with duplexing and switching are not shown in this simplified block diagram.

[21]SAW filters typically have very good selectivity but suffer from relatively high losses.

[22]In some cases the signal is translated to a low IF center frequency rather than DC

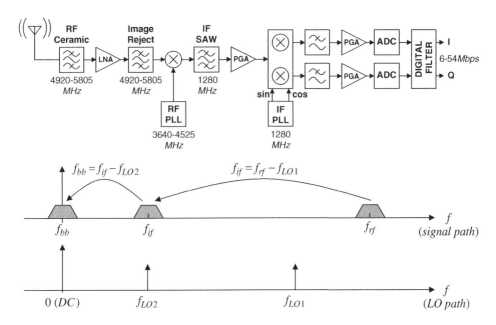

Figure 2.1 (**a**) Block diagram of "typical" superheterodyne 802.11a receiver. (**b**) Frequency translation in superheterodyne receiver (negative frequencies and image rejection not shown).

the desired signal range of the ADC. The signal is then sampled by the ADC and passed on to the digital baseband for further processing.

It is worth discussing some important aspects of the selection and use of the filters in the superheterodyne architecture in more detail:

- The *band-select filter* acts to prevent the desensitization and possible saturation of the LNA. Often the LNAs are designed with a tuned input matching and load impedance. In such cases and in a relatively interference free environment, the band-select filter may be eliminated. When the band-select filter is to be used, the designer needs to be aware of the direct impact of this filter on the noise figure of the system (since there is no gain before this filter). Typical band-select filters for WLAN have a loss of 1.5 to 3 dB, but this number can vary significantly depending on the specific requirements and the amount of selectivity required. Note that in some cases the image-reject filter is eliminated.[23] In such cases, the band-select filter would also act to

---

[23]This filter may be eliminated for a variety of reasons, including to save cost or area or to avoid having to go off chip and back on chip again.

reject any signals present at the image frequency (more on the concept of image below).

- The *image-reject filter* is primarily responsible for eliminating any signal or noise present at the image frequency. A question that may arise is that if the band-select filter is present in the system and acts to reject any interferers at the image frequency, why would a second dedicated image-reject filter be required? The answer to this question will become apparent when one realizes that the LNA may have a fairly broadband noise output and therefore have a significant amount of noise at the image frequency. It can easily be shown that, if the following assumptions hold, then, the noise figure (NF) of the system will increase by 3 dB as compared to a case where a lossless image-reject filter is used:

  1. The noise at the output of the LNA has a white spectral density (i.e., it is flat versus frequency).
  2. No image-reject filter is used (even if a band-select filter is used).
  3. The noise is dominated by the LNA and the components that precede it.

  Typically, the assumptions above are only partially true, and a real image-reject filter is not lossless. As a result, eliminating an image-reject filter would degrade noise figure by an amount less than 3 dB.[24]

- If a high linearity LNA can be used in the system, typically the band-select filter is eliminated. In such cases the image-reject filter would act as the band-select filter also, but with much less degradation of NF (since it comes after the LNA, which has gain).

- The channel-select filter (in this case the SAW) has a high degree of selectivity. In the case of a WLAN system, such a filter is capable of selecting the desired channel only. This is because the channel bandwidth is fairly high. For example for 802.11g, the SAW filter would be required to have a Q of approximately 1280 MHz/16 MHz = 80. In some applications, however, selecting one channel at the IF may not be possible due to the high Q required.[25] In such cases further low pass filtering at the baseband frequency would be required.

---

[24]This concept is also the source of the terms single-sideband (SSB) noise figure versus double-sideband (DSB) noise figure. In a typical mixer, the DSB NF is 3 dB less than its SSB NF. It is also important to note that in cascaded NF calculations, the SSB NF should be used for superheterodyne and low IF architectures, and DSB NF should be used for direct-conversion receivers. This fact should become clear upon the completion of the receiver architecture discussions.

[25]This would clearly depend of the bandwidth of the channel as well as the selected IF.

- The baseband filters may be required to supplement the action of the IF channel-select filter. Even if the IF channel-select filter can exclusively select the desired channel, the baseband low pass filters may be required to filter out the broadband noise present at the output of the baseband components. Without such filtering, the ADC could alias the noise at the out-of-band frequencies and degrade the SNR.

- Even in the presence of the band-select and image-reject filters, the RF components that precede the channel-select filter need to have a high degree of linearity in order to be able to deal with the existence of multiple channels and therefore large signal swings at their input.

2.1.1.1 Choice of Intermediate Frequency in Superheterodyne Receiver

An image signal is one that would fall on top of the desired signal after the mixer's frequency translation action. As an example, for the superheterodyne receiver of Figure 2.1, in order to tune into the 4920-MHz band, and with an IF of 1280 MHz as shown, the RF PLL would have to tune to 3640 MHz (since 4920 MHz − 1280 MHz = 3640 MHz). In this case the image frequency would fall at $F_{im} = F_{rf} - 2F_{if} = 2360$ MHz. It is important to note that if the image signal (or noise) is not properly filtered or otherwise rejected using an image-reject mixer, it would mix with the desired signal and it would not be possible to remove it using any postprocessing.

It should also be clear that the choice of the IF in a superheterodyne architecture is a very key decision. This decision should be made with many considerations, including the following:

- What is the power level of a potential interferer in the image band? The relative strength of the interferer as compared to the minimum desired signal is quite important and should be kept to a minimum.

- What type of interferer is present at the image band (is it CW, spread spectrum, etc.)? Certain kinds of modulated signals have more sensitivity to specific types of interferers.

- Is the desired incoming band of signals broadband in nature? Choosing the proper IF for a broadband input signal is more challenging, since the designer needs to make sure than the image for none of the desired channels would cause problems. In general, the IF needs to be selected to have a frequency larger than the difference between the highest desired channel frequency and the lowest desired channel frequency (assuming a contiguous band of desired signals). In our example, the IF needs to be selected to have a frequency greater than 5805 MHz − 4920 MHz = 885 MHz. If this criterion is not observed, one of the channels in the band of interest would be the image for another channel. In our example, had we chosen the IF to be 300 MHz and we wanted to select the 5180-MHz channel, the image frequency would have fallen at 5780 MHz. Since the band-select filter and the image-reject filter need to allow the entire band to pass through, they would have not rejected the image, and the image would have fallen in the desired band. Note that, often, the image frequency is selected as a lower frequency than the RF in order to simplify signal processing and reduce power consumption.[26]

- Should the system be designed with a high side LO or a low side LO? The answer lies in the presence of large interfering image signals with each choice.

- How high should the IF be? In general, the higher the IF, the easier for the band-select and image-reject filters to reject the image. However, at the same time, it would become more difficult for the IF band-select filter to select the desired channel (a higher Q would be required for a given channel bandwidth).

There are several other important factors than enter the equation. These include issues such as the availability and cost of commercial filters at the desired band[27] and the presence of large interferences at the IF (which can couple directly to the IF stage).[28]

By now it should be clear that the choice of the IF for a superheterodyne receiver requires consideration of many factors and should not be taken lightly.

It is important to note that several variants to the basic superheterodyne architecture have been proposed and are in use. One such architecture is the "sliding IF" architecture as shown in Figure 2.2. In this architecture, instead of a fixed IF, a sliding IF is chosen. The IF is therefore generated by dividing the RF LO frequency by an integer multiple. The IF therefore changes

[26]One exception to this rule is the case of cable tuners. The desired channel can range from approximately 50 to 850 MHz. Clearly there are no suitable IF present below 50 MHz. In such cases an "up–down" approach is often used where the first IF has a frequency higher than the incoming frequency.

[27]If the IF filter is to be designed from off- or on-chip inductors and capacitors, the queue of the available components at the IF and their potential size should also be considered.

[28]It should be noted that typically tunable high Q front-end filters are not available. (Note: "Q" is an abbreviation for quality and is a measure of the selectivity of the filter.) Therefore the discussions above center around fixed-frequency RF filters. The presence of tunable high Q filters would significantly impact the choice of radio architectures and simplify the design of the active blocks that follow such a filter. An area of active research which holds promise for possibly enabling tunable high Q filters in the future is microelectromechanical systems (MEMS).

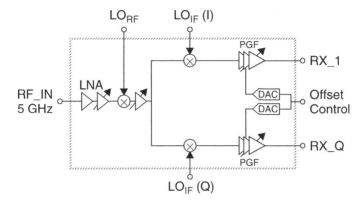

Figure 2.2 Block diagram of "sliding IF" superheterodne 802.11a receiver. The frequency of the RF LO is an integer multiple of the IF LO. This architecture avoids the need for two PLLs for the superheterodyne receiver (Su et al., 2002).

as a function of the selected channel (hence the term sliding IF). In other words, for any given selected channel, the RF LO is operating at a frequency that is harmonically related to the frequency of the IF LO. This technique eliminates the need for a second (IF) PLL but makes channel selection at the IF more difficult. As a result most of the channel selection would have to be implemented using baseband filters.

The following highlights the advantages and disadvantages of the superheterodyne architecture as compared to the other receiver architectures:

- On the positive side the superheterodyne architecture:
  1. Is well known and well understood. It provides a reasonably good performance. Implementing a superheterodyne architecture is relatively low risk.
  2. Can achieve a reasonably low power consumption.
  3. Allows a reasonably flexible frequency plan (choice of the IF is somewhat flexible subject to the constraints discussed above).
  4. Avoids DC offset problems associated with the direct conversion and in some cases the low IF architectures. There are three primary reasons:
     (a) There is a significant amount of gain before the signal hits the baseband section. Therefore, from a gain budgeting point of view, less gain at baseband would be required. This reduces the potential issues associated with gaining up offsets associated with device mismatches and potentially saturating the following stages.

(b) The self-mixing is typically not an issue. Self-mixing is a process in which the LO signal can couple to the RF port and mix with itself causing a *time-variant DC offset*. In a super-heterodyne architecture the RF LO self-mixing is out of the band of interest for the desired signal and is easily rejected by simple high pass filtering [alternating current (AC) coupling as an example]. On the other hand, the IF LO self-mixing does create DC offsets that fall in the band of interest. However, as compared with the other architectures, this LO is running at a relatively low frequency and therefore the amount of coupling from the LO port to the RF port would be significantly less in magnitude, resulting in smaller amount of DC offsets at baseband.

(c) Due to the IF stage filtering, the final down-conversion mixer does not see large interfering signals, which can result in large DC offsets at baseband. DC offsets generated due to large interferers by the components before the IF filter will typically be rejected by the AC-coupling stages utilized in such stages.

5. Avoids issues associated with the even-order harmonics as a result of large incoming interferers at the antenna. Some of these issues are DC related as described in the previous paragraph.

6. Achieves superior quadrature balance as compared to the alternative architectures. This is primarily due to the fact that the quadrature signals for this architecture are generated at the IF rather than the RF. Therefore the impact of parasitic mismatches in the generation and distribution of the in-phase and quadrature-phase signals are less. To the first order and assuming no auto-IQ calibration is used, if one can generate an IQ signal with a phase accuracy (standard deviation) of 2 degrees at 5 GHz, they would be able to generate a phase accuracy of approximately 0.5 degrees at 1.25 GHz. Better quadrature matching is also achieved due to the fact that the signal shares a common path over a larger portion of the receiver chain as compared to a direct-conversion or low IF receiver.

- On the negative side, utilizing a superheterodyne architecture:

1. Results in a design that is expensive and large as compared to the alternative architectures.

2. Requires many discrete and external components. This shortcoming and the previous one are the primary reason for designers moving away from this architecture for integrated solutions.

3. There is an image problem to be reckoned with. Since the image frequency is typically far away from the desired signal, it often

falls at a frequency which is not even regulated by the same standard and it may therefore be more difficult to design for. Further there may be interferers at certain parts of the world but not others. Therefore a design that is intended for the global market should investigate the frequency allocations for various countries.

4. In general, it would be difficult to design a multimode superheterodyne receiver that performs the channel selectivity at the IF. For a multimode receiver, the channel bandwidth and selectivity would need to be programmable. This is not possible to do with single discrete filters such as SAWs. It is also difficult to do a reasonable job on silicon due to the limited queue of the passive devices at the IF. The alternative would be to use multiple parallel filters and use the proper one depending on the required mode of operation. However, this would be quite expensive and bulky. Alternatively, in the case of a multimode superheterodyne receiver, the channel selection would therefore have to be performed at baseband with programmable baseband filters (opamp-RC or Gm-C). This takes away from some of the advantages of the superheterodyne architecture.

5. As discussed in the section on the choice of the IF, finding a suitable IF in the case of a broadband input may be challenging.

It is important to note that the above disadvantages are general statements. Many of the disadvantages of the superheterodyne architecture can be overcome (for certain applications and with certain trade-offs) with smart architectural and circuit design techniques. However, the general statements hold and are valid.

2.1.2 Low IF Receiver

An alternative architecture to the superheterodyne is the low IF. This architecture is fundamentally similar to the superheterodyne approach. However, as the name suggests, a low (i.e., close to baseband) frequency is chosen for the IF. As a result of this choice, some fundamental changes are required in the way the receiver is assembled. The low IF approach can be viewed as an architecture with characteristics somewhere between a superheterodyne and direct conversion. Therefore one would expect that the low IF architecture would share some of the advantages and disadvantages of the superheterodyne and direct-conversion approaches.

One fundamental difference in the lineup arises from the fact that due to the low IF, a (fixed-frequency) front-end RF filter would be incapable of rejecting any interferers present at the image frequency. As a result, image-rejecting mixers or filters would have to be used instead. Also note that the choice of the low IF is not arbitrary. The low IF has to be chosen large enough in order to avoid aliasing of the signal at baseband. This requires the low IF to be *at least* half the bandwidth of the RF. The choice of the IF for the example of Figure 2.3 was set based on this criterion. More discussion on this topic will follow later.

As shown in Figure 2.3, in the low IF approach (illustrated for a hypothetical 802.11a system), the signal is received from the antenna and passed on to a channel-select filter.[29] The signal is then amplified by an LNA before being sent to quadrature mixers. The in-phase and quadrature-phase outputs of the mixers are then passed on to low pass baseband filters for channel selection. The output of the filters are passed on to baseband PGAs and then sent to the ADCs. The final down conversion to baseband and actual image rejection are performed in the digital domain. Because of this fact, the entire baseband analog blocks as well as the ADC would have to have a large enough dynamic range (more that would otherwise be required) to be able to handle the potentially large image signal that may be present at their inputs. This requirement for a higher dynamic range typically translates into more power consumption.

In an alternative low IF implementation, as shown in Figure 2.4, the signal out of the RF quadrature mixers is passed on to complex bandpass filters. These filters act to reject the image, before passing on the signal to the remainder of the baseband analog circuits and the ADC. This reduces the dynamic range requirements (and therefore power consumption) on these blocks but at the same time increases the complexity and the power consumption of the filters:

- On the positive side, the low IF architecture:
  1. Eliminates the IF SAW, IF PLL, and image-filtering components of the superheterodyne architecture.
  2. Allows for a high order of integration.
  3. Relaxes the image rejection requirements as compared to a superheterodyne architecture. This is because in a low IF architecture the channel present at the image frequency is likely under the regulation of the same standard as the desired channel. As such, often, strict guidelines exist on the maximum power level that the (closely spaced) image channel will have. In contrast, in a superheterodyne architecture, with its high IF, the image frequency can often have a

---

[29]Blocks associated with duplexing and switching are not shown in this simplified diagram. Also note a balun is not explicitly shown but assumed in the block diagram before the LNA. This balun converts the single-ended signal to a differential signal.

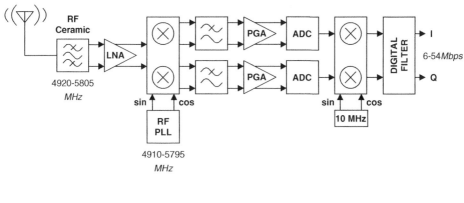

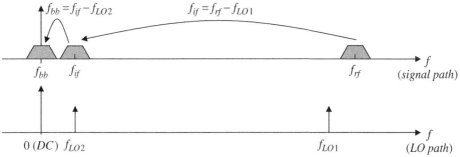

Figure 2.3 (**a**) Block diagram of low IF 802.11a receiver. The low IF frequency has to be chosen to accommodate various trade-offs. In this implementation of low IF receiver the burden of image rejection is passed on to the digital mixers and filters. (**b**) Frequency translation in low IF receiver (negative frequencies and image rejection not shown).

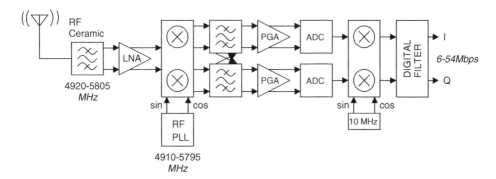

Figure 2.4 Block diagram of another low IF 802.11a receiver. In this version of the implementation of the low IF receiver the burden of image rejection is primarily handled in the analog domain. Note the complex bandpass analog filters compared to the simple low pass analog filters of the alternative low IF implementation of Figure 2.3.

very strong amplitude level. No matter how strong the image component amplitude is, the receiver should be able to operate properly.

4. Reduces DC offset problems associated with the direct-conversion architecture. However, it is not as robust as a superheterodyne architecture in regards to DC offsets. Unlike a superheterodyne architecture, there is limited amount of gain at the RF stages before reaching the low IF section. As such, the DC offsets at the low IF, whether due to self-mixing or mismatches, will be amplified by the large gain present at the low IF chain. On the other hand, canceling the generated DC offsets using high pass filters (e.g., AC coupling) is easier than an equivalent direct-conversion system. This is because the low IF is chosen such that there is minimal modulation information content close to DC (in the direct-conversion scheme and ignoring the effect of frequency offsets due to the crystals, "DC" would be exactly in the middle of the band).

- On the negative side, utilizing a low IF architecture:

1. Requires the generation of the quadrature signals at the RF. This is more difficult to do than at the lower frequencies in which the superheterodyne architecture would generate its quadrature signals.

2. Requires higher performance (and hence power hungry) ADC and/or high dynamic range (and therefore power hungry) bandpass filters.

3. An additional set of second (low frequency) mixers are required. These mixers may be implemented in the digital domain (but would require the higher performance ADCs).

4. Flicker noise may be a problem, especially in complementary metal–oxide–semiconductor (CMOS) implementations. This is primarily due to the fact that the signal has not been significantly amplified before reaching the high flicker noise region of the receiver (i.e., the low IF components). Clearly flicker noise is much more of an issue for low IF than for superheterodyne but less so for low IF than direct conversion.

5. In certain applications [such as cellular Global System for Mobile communications (GSM)], the choice of a low IF (often 100 kHz) may cause a slow RF PLL settling in the system if an integer-N PLL is used. This is because in an integer-N PLL, the lowest frequency that is to be resolved is directly related to the loop bandwidth of the PLL. So to resolve a low IF, the loop bandwidth (BW) of the RF PLL may need to be reduced. This may have an undesirable impact on the settling behavior of the RF PLL.

So is a low IF architecture suitable for a WLAN system? There have been publications using a low IF approach for WLAN receivers; however, this architecture is not the most popular one for WLAN. The primary reason for this is that the wide bandwidth for the WLAN modulation causes the IF for the low IF architecture to be relatively high. As a result the power consumption of such a solution is typically higher than an equivalent direct-conversion architecture. We will discuss the reasons for this in more detail after introducing the direct-conversion architecture.

Note that the low IF architecture is a very popular choice for narrowband modulated signals such as Bluetooth (with typical low IF of 2 MHz) and GSM (with typical low IF of 100 kHz).

2.1.3 Direct-Conversion Receiver

Another alternative to the superheterodyne architecture is the direct-conversion or zero-IF architecture. As shown in Figure 2.5b (show spectrum translation) and ignoring frequency inaccuracy issues with the reference (e.g.,

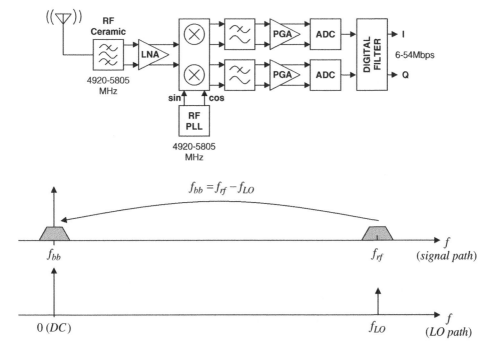

Figure 2.5 (**a**) Block diagram of direct-conversion (or zero IF) 802.11a receiver. (**b**) Frequency translation in direct-conversion receiver (negative frequencies not shown).

crystal oscillator), the RF-modulated signal is directly translated to baseband (i.e., zero IF). As seen in this figure, after down conversion, "DC" falls in the middle of the modulated signal. This has consequences which will be discussed shortly.

An example of such an architecture applied to the 802.11a standard is shown in Figure 2.5. Note that the signal flow in this architecture is very similar to that shown in Figure 2.3 for a low IF receiver. In this example, the signal received from the antenna is passed through a band-select filter.[30] Then the signal is amplified by the LNA and passed on to the quadrature mixers. The frequency of the RF PLL driving the quadrature mixers is selected in such a way that the down-converted signal would fall directly at baseband (i.e., $f_{LO} = f_{RF}$). In this example the down-converted signal is low pass filtered and amplified by the PGAs before being passed to the ADCs.

The idea of direct-conversion receivers has been around for a very long time. In practice, however, the earlier generation designs were often plagued with various problems. As a result, until several years ago, not many production direct-conversion receivers were in commercial use. In contrast, the majority of WLAN transceivers shipped today utilize a direct-conversion receiver. There are several techniques utilized in today's receivers which enable the successful use of a direct-conversion architecture. Given the popularity of this architecture for WLAN designs, the remedies for the shortcomings of the direct-conversion architecture are described below.

The advantages and disadvantages of a direct-conversion receiver are outlined as follows:

- On the positive side, the direct-conversion architecture:
  1. Similar to the low IF architecture, eliminates the IF SAW, IF PLL, and image-filtering components of the superheterodyne architecture.
  2. Allows for the highest level of integration.
  3. Offers the lowest cost. This is probably the strongest reason in favor of adopting a direct-conversion architecture (assuming that all challenges associated with the architecture as outlined below can be worked out).
  4. "Avoids" the image problem. This is because the "image" is the signal itself. Therefore, at any given time the power level of the image is no more than the power level of the desired signal. Further the image has information content associated with the desired sig-

---

[30]Blocks associated with duplexing and switching are not shown in this simplified diagram. Also note that a balun is not explicitly shown but assumed in the block diagram before the LNA. This balun converts the single-ended signal to a differential signal.

nal. It is important to note that accurate quadrature balance is still important in a direct-conversion receiver, so that proper EVM levels can be maintained. More on this topic later.

5. Can be low power. Some early examples of direct-conversion receivers were higher power that their superheterodyne counterparts. However, recent examples of WLAN receivers using direct-conversion architectures show some of the lowest power consumptions reported using any architecture.

6. Can be high performance. Earlier designs with direct-conversion architecture were not as high performance as the superheterodyne ones. However, recent publications of WLAN receivers describe direct-conversion receivers with excellent performance numbers.

• On the negative side, utilizing a direct-conversion architecture:

1. Requires the generation of the quadrature signals at the RF. This is more difficult to do than at the lower frequencies in which the superheterodyne architecture would generate its quadrature signals. Utilizing proper matching design and layout techniques as well as possibly some form of quadrature calibration allows the designers to overcome this shortcoming.

2. If LO reradiation is of concern,[31] the LO generation circuit can be designed in such a way that the VCO runs at an offset (or harmonic or subharmonic) frequency relative to the incoming RF signal. This strategy does not completely eliminate the LO reradiation problem, but it reduces the mechanisms by which it could occur. For example, with a brute-force approach such as that of Figure 2.5, the LO frequency is exactly the same as the desired frequency (e.g., 5240 MHz) and can reradiate out through direct VCO coupling to the LNA inputs. However, if the LO is synthesized using a "harmonic" approach, the VCO can run at 2620 MHz, and it can then generate the required 5240 MHz after the VCO stage.

3. Even more so than the low IF architecture, flicker noise can be a problem, especially in CMOS implementations. This is primarily due to the fact that the signal has not been significantly amplified before encountering the high flicker noise region of the receiver (i.e., the baseband section). Fortunately in the case of a WLAN

---

[31]LO reradiation is the process by which the LO frequency which falls at the center of the incoming received signal for a direct-conversion receiver reradiates (transmits) back through the antenna. LO reradiation can be caused by several phenomena such as asymmetric layout, poor reverse isolation of RF blocks, excessive LO-to-RF coupling paths, and coupling of the VCO signal through the package to the LNA inputs (if $f_{VCO} = f_{LO}$).

system, the wideband nature of the modulated signal reduces the flicker noise problem. This is because the flicker noise primarily impacts the inner subcarriers of the modulated signal, and for a good design, the higher frequency subcarriers are often dominated by thermal noise. In contrast, an otherwise equivalent narrow-bandwidth-modulated signal down converted to baseband would be impacted by flicker noise *on average* more significantly. This is one reason why CMOS Bluetooth and GSM-based receivers seldom use a direct-conversion architecture.

4. It can be significantly impacted by DC offsets. Unlike a super-heterodyne architecture and similar to the low IF architecture, there is a limited amount of gain at the RF stages before hitting the low IF section. As such, the DC offsets at the baseband frequency, whether due to self-mixing or mismatches, will be amplified by the large gain present at the baseband chain (60 to 80 dB of baseband gain for typical direct-conversion WLAN receivers is not uncommon). On the other hand, canceling the generated DC offsets using high pass filters (e.g., AC coupling) is more difficult than an equivalent direct-conversion system. This is because, in the direct-conversion scheme and ignoring the effect of frequency offsets due to the crystals, DC would be exactly in the middle of the band. Therefore, eliminating the DC components using any kind of high pass filtering has the potential of cutting off information content included at the center of the baseband-demodulated channel. The good news is that in the case of the WLAN OFDM signal the zero-order subcarrier (which would fall exactly at DC in a direct-conversion receiver) is eliminated in the standard exactly for this reason. Even then, the bandwidth of the high pass-filter(s) (HPFs) should be selected carefully. Too low of a HPF corner would causes slow transient response. Too high of a HPF corner could cause eating into the lower order subcarriers. The wide bandwidth of the WLAN signal helps in this case the same way that it does with the flicker noise problem. For WLAN receivers, smart schemes have been devised and published that allow for an excellent performance despite the potentially large DC offsets at baseband.

5. Even-order nonlinearities and amplitude-modulated signals can cause problems in a direct-conversion receiver. Filtering in the transmitter, channel impairments, or the modulation itself can cause the received signal to include AM. This AM, in conjunction with the limited second-order intermodulation performance of the RF front end, can be translated into baseband and cause the corruption and EVM

degradation of the received baseband signal. This phenomenon is typically called *AM detection*. AM detection can also be caused by even-order nonlinearities in the baseband blocks themselves.[32]

6. Even-order nonlinearities can cause problems in a direct-conversion receiver in an additional way. Unlike a superheterodyne receiver where the IM2 component $(f_1 - f_2)$ of the RF signal would fall out of band, in this case these two components can mix with each other and at least partially fall in the band of interest and degrade the performance of the receiver. Therefore the design of a RF front end with high IIP2 is desirable. It is important to note that narrowband interferers in this case can cause an additional (undesirable and time-varying) DC signal which can be fairly easily rejected by the HPFs. However, broadband interferers can create broadband baseband interference that can interfere with the desired signal. Normally the IIP2 of a receiver is dominated by the mixer (due to the high gain of the LNA and AC coupling between LNA output and mixer output). However, there are interactions between the DC offsets and the even-order nonlinearities that can cause the baseband blocks to contribute to this problem. Assume that a baseband amplifier with infinite IIP2 can be designed (clearly not possible in reality).[33] Such an amplifier would still possess DC offsets at its input and finite odd-order nonlinearities. By going through the power series expansion of such an amplifier, one can see that even-order nonlinearities are generated. The magnitude of the even-order nonlinearities would be a function of the amount of DC offsets and the magnitude of the odd-order nonlinearities in the amplifier.

2.1.4 Receiver Architectures: Summary

So what architecture should one utilize? There is no ubiquitous answer to this question. The choice of the proper receiver architecture is dependent on many factors, including the required system level trade-offs, modulation type, background and expertise of the designers with the various architectures, and risk tolerance.

---

[32]Although the baseband signals typically have a high second-order intercept point (IIP2), they are preceded with the LNA and mixer at the front end, which amplifies the signal significantly. This large signal along with the finite IIP2 of the receiver baseband blocks can then cause AM detection.

[33]The IIP2 of a baseband amplifier can be made to be very high by using symmetric layout techniques and large-size devices. These same techniques usually do not apply to high frequency blocks due to the excessive parasitic capacitances they create.

A general trend, however, is the move to direct-conversion and low IF receivers and away from superheterodyne architectures. This trend is driven by the push for low cost, high integration designs with minimal external components. In the case of WLAN, the majority of the designs are of the direct-conversion type due to the wideband nature of the modulation used.

Table 2.1 summarizes the advantages and disadvantages of the various receiver architectures that have been discussed in this chapter.

2.1.5 Superheterodyne Transmitter

Now that we have discussed the various receiver architectures, we need to understand the choice of the transmitter architectures. The duals of many of the advantages and disadvantages that apply to the receiver architecture do apply to the transmitter. Here we will present a summary of these advantages and disadvantages and only discuss in more detail cases which are specific to the transmitter.

Figure 2.6 display a typical superheterodyne transmitter. In this transceiver the baseband digital bits are received from the digital PHY block and

Table 2.1 Summary of Advantages and Disadvantages of Various Receiver Architectures Applied to WLAN

Receiver Architecture	Advantages	Disadvantages
Superheterodyne	Good overall performance Usually low power Flexible frequency plan Avoids DC problems	Expensive, large Many discrete, external components Image problem Difficult for multimode (need multiple IF filters)
Direct conversion	Lowest cost *Can* achieve very good performance and low power for WLAN Eliminates IF SAW, IF PLL, and image filtering Integration Avoids image problem	Quadrature RF down conversion required: DC problems; AM suppression, LO self-mixing, DC offsets and $1/f$ noise (CMOS) Typically requires offset or $2 \times$ LO to avoid coupling
Low IF	Eliminates IF SAW, IF PLL, and image filtering Avoids DC problems Integration	Quadrature RF down conversion required Requires higher performance ADC or BB filters May require offset LOGEN Typically higher power consumption for WLAN applications

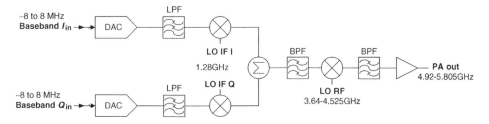

Figure 2.6 Block diagram of superheterodyne 802.11a transmitter.

passed on through the DACs to generate the appropriate I and Q analog baseband signals. The signals are then passed through the proper low pass filters. These filters act to reject any high frequency aliasing caused by the DAC that may act to violate the spectral mask of the standard. The orders of these filters are set by spectral mask requirements, the oversampling rate of the DAC, and the characteristics of the digital filters preceding the DACs.

In this example the filtered I and Q signals are passed on to quadrature up-converting mixers running with LO signals at 1.28 GHz and then combined together to generate the single-sideband-modulated signal at the IF (1.28 GHz). The signal is then typically passed through a bandpass filter which acts to reduce the spurs and can reject any residual DAC aliasing that may have not been completely rejected by the digital and analog baseband filters. The IF filtered signal is then passed on through the RF mixer which generated the RF signal at 4.9 to 5.805 GHz depending on its LO frequency. Depending on the application another stage of filtering may be applied here before passing the signal through a power amplifier or power amplifier driver.

One reason for the need for so many filters in this architecture is the multiple up conversions. Every time there is a mixing action, one needs to be aware of any image components that may be created (this image component may be an image signal or, as is frequently the case, image band noise). If the in-band noise or out-of-band noise of the transmitter is of importance,[34] it would be critical to reject the image noise prior to up conversion. Otherwise due to a reason similar to the phenomena described in the superheterodyne receiver case, the transmitter SNR can suffer by as much as 3 dB.

---

[34]The in-band noise of the transmitter would be of importance if the transmitter performance is being set by the SNR of the transmitter. This is rarely the case. However, the out-of-band noise of the transmitter may be of importance due to spectral regrowth and spectral mask issues. Further the out-of-band noise of a transmitter is almost always of importance in full-duplex applications where the transmitter leakage signal into the receive band can desensitize the receiver. The out-of-band noise of the transmitter is also of importance in multiradio applications (e.g., WLAN in a WCDMA cell phone).

One very important advantage of a superheterodyne architecture as compared to a "brute-force" low IF and especially direct-conversion architecture arises from the fact that the VCO frequencies are such that they inherently are not operating near the operating frequency of the power amplifier. This is important because if the VCO operates at or near the frequency of the power amplifier the VCO can be "pulled" (more on this later), causing a significant degradation in transmitted signal quality. The pulling effect typically becomes more of an issue at higher frequencies (where it is difficult to provide sufficient isolation) and higher PA output levels.

Another advantage of the superheterodyne architecture is that it eliminates the LOFT signal in the output caused by the direct coupling of the LO to the RF (this issue will be discussed in more detail later).

It is important to note that, although the superheterodyne architecture has a natural immunity to VCO pulling, the choice of the IF can impact the degree of resilience of the VCO to the PA pulling. Ideally, the VCO frequencies should be chosen such that the PA output frequency is not harmonically related to the VCO.

So, in summary, the superheterodyne transmitter has a good performance, is reasonably low power,[35] offers flexible frequency planning, can achieve a very good quadrature balance, can reject extraneous spurs due to extended filtering, and has a high degree of resilience to PA pulling.

On the other hand, a superheterodyne transmitter is comparatively expensive and large, requires many discrete components, is not very suitable for multimode applications due to the narrowband nature of the IF filter, and provides restricted IF choices for applications with broadband transmitter output requirements.

An example of a modified superheterodyne transmitter architecture is shown in Figure 2.7 (Su et al., 2002). In this architecture, the quadrature baseband signals are applied to a bank of quadrature up-converting mixers which up convert the signal to the IF. The signal is not combined at the IF but rather is amplified by variable-gain amplifiers as independent quadrature signals. The I and Q signals are then applied to a pair of quadrature RF mixers which up convert the signals to the desired RF before combining them to generate the image-rejected output signal. This combined signal is then applied to the RF power amplifier or power amplifier driver block. There are several differences between this method and the traditional superheterodyne architecture that are worth mentioning.

First, as can be seen, the external and bulky filters have been eliminated, allowing for a much more compact implementation of this architecture as

[35]The fact that this architecture requires driving off chip filters, typically with low input and output impedances, increases the required power consumption, however.

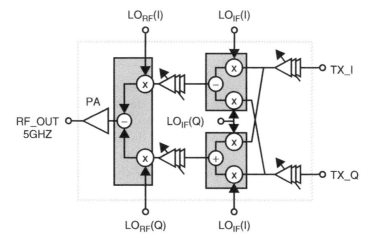

Figure 2.7 Block diagram of "sliding IF" superheterodyne 802.11a transmitter. Similar to the receiver, the frequency of the RF LO is an integer multiple of the IF LO (Su et al., 2002).

compared to a traditional superheterodyne architecture. The image issues are resolved by utilizing image-reject mixers and maintaining the signals in quadrature form baseband all the way to the desired RF. The accuracy of the single-sideband-modulated signal (and therefore the transmitter image rejection) in this architecture is primarily set by the first up-conversion quadrature mixers. Since these mixers operate at much lower frequency than the RF, very good transmit image rejection can be achieved. The second set of image-reject mixers allow for the elimination of the IF filtering of the sideband.

Second, this architecture eliminates the need for an independent PLL for the IF stage. The RF is selected to operate at an integer multiple of the IF. As such, the PLL generates the required LO RF signal, and a divided version of that signal is used as the low IF.

As can be seen, this architecture eliminates some of the shortcomings of the traditional superheterodyne architecture while maintaining some of the advantages of that architecture. However, with everything else being the same, it would be larger than an equivalent direct-conversion transmitter with its six up-converted mixers and associated LO buffers (as compared to two mixers with their associated buffers for the direct-conversion transmitter).[36] Further this architecture would likely burn more power than the

---

[36]In a direct-conversion transmitter in which the VCO is designed to operate at a different frequency than the PA, additional mixers are typically utilized. These mixers add to the area and power consumption of such a direct-conversion architecture. However, they are not in the signal path and, if designed properly, do not contribute to any significant EVM degradation.

equivalent direct-conversion transmitter due to its need for the typically power hungry LO buffers.

2.1.6   Low IF Transmitter

The dual for the low IF receiver is the low IF transmitter. The low IF transmitter operates very similarly to a direct-conversion transmitter but requires full image-reject mixers. As compared to the direct-conversion transmitter, it has the advantage that DC baseband offsets do not translate into an in-band LOFT (although they could create out-of-channel LOFT which may violate spectral mask requirements and may be difficult to filter). Further, the RF output for a low IF transmitter operates slightly offset from the output frequency of the VCO, and this provides some (but typically not much) resilience to VCO pulling. In practice, in most cases, the complexity associated with the low IF transmitter is not worth its advantages and is seldom used for WLAN applications.

Low IF transmitter architectures are more popular for narrowband signals. Often, when low IF architectures are utilized, the first stage of up conversion is performed in the digital domain such that a high degree of image rejection can be maintained.

2.1.7   Direct-Conversion Transmitter

An example of a direct-conversion transmitter is shown in Figure 2.8. In this example, the baseband I and Q signals are applied to the DACs and passed through the antialiasing filters (recall that these filters help satisfy the spectral mask requirements as explained in the superheterodyne transmitter section). The filtered signals are then applied to quadrature up-converting mix-

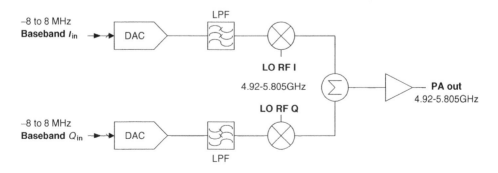

Figure 2.8   Block diagram of direct-conversion transmitter.

ers before being combined and being applied to the power amplifier or power amplifier driver.

As the dual of the direct-conversion receiver, this transmitter offers the lowest overall cost and eliminates the IF bandpass filter, the IF PLL, and the requirements for image filtering. As such it offers the highest degree of integration. This architecture can be made to have low power consumption and high performance.

On the other hand, this architecture requires a quadrature RF up conversion of the baseband signals. As described in Section 2.1.3, obtaining high accuracy quadrature LOs running at the high RF is difficult.

Probably the biggest issue with a traditional direct-conversion transmitter is the fact that the VCO operates at the same exact frequency as the power amplifier (or power amplifier driver). The power amplifier can couple to the VCO by various means (through bond wires of the package, through the substrate, through the finite reverse isolation of the LO buffers, etc.) and pull the VCO, causing a degradation in modulated signal quality.[37]

The solution to the VCO pulling inherent in a direct-conversion transmitter is the use of an offset LO frequency. In this mechanism, the LO is generated at a different frequency than the RF (i.e., the VCO runs at a different frequency than the PA). The desired RF is then generated by multiplying or dividing the generated LO by an integer or fractional number (e.g., 2, $\frac{1}{2}$, $\frac{2}{3}$). From a VCO-pulling point of view, the best solutions are those in which the VCO frequency and the PA frequency are not harmonically related (e.g., $F_{vco} = \frac{2}{3} F_{rf}$). Of course it can be argued that even in this scheme, the second harmonic of the PA falls at the same frequency as the third harmonic of the VCO; however, the amount of coupling in these cases is typically significantly smaller than the other cases.

Note that using $f_{LO} \neq f_{RF}$ does require some additional signal processing on the LO signal and would therefore add some size and power consumption to the solution. However, given the risk of potential VCO pulling, it is indeed rare to find a direct-conversion transmitter design in which $f_{RF} = f_{LO}$.

Another typical problem with the direct-conversion transmitter is the DC offsets contributing to the LOFT in the middle of the desired transmitted band. Various standards have certain requirements on the maximum amount of LOFT allowed, and these requirements need to be met. In a direct-conversion transmitter, DC offsets of the baseband blocks can generate

---

[37]VCO pulling is fairly easy to detect and verify in the lab by performing a set of experiments (including reducing the transmitter output power by various means to identify the location and mechanism for coupling).

LOFT. Further the LO can couple at the RF directly to the output signal and create LOFT. The ways to work around the LOFT problems in a direct-conversion transmitter is the use of good layout techniques as well as LOFT calibration algorithms.

In summary, similar to the receiver case, the choice of the transmitter architecture is a function of many factors. These include the degree of risk tolerance required, the experience of the designers on the various architectures, modulation type and characteristics, package type, substrate resistivity, whether or not a power amplifier is being integrated, the choice of process technology, and the desired transmitter output power level. It is important to note that for time-duplexed systems (like WLAN) it is quite beneficial to use the same architecture for both the receiver and the transmitter; otherwise two different LO frequencies would be required during transmit and receive. Therefore, either two PLLs or a fast-settling PLL would be required.

2.1.8 Polar Modulators

The transmitter architectures discussed in previous sections utilize quadrature (or Cartesian) inputs. In other words, any point on the constellation diagram (which can be represented by a vector in the signal space) is represented by the magnitude of the I and Q signals (Fig. 2.9). In this class of transmitters, the signal information at baseband is contained in I and Q inputs, with each of the rails utilizing its own mixer. The up-converted and Q signals are combined at IF or RF to generate the modulated signal.

The constellation points (and therefore modulated data) can also be represented utilizing polar coordinates as shown in Figure 2.9. In the signal space, the magnitude component of the signal, β, is given by

$$\|\beta\| = \sqrt{I^2 + Q^2}$$

And the phase component of the signal is given by

$$\alpha = \tan^{-1}\left(\frac{I}{Q}\right)$$

A transmitter that generates the RF signal utilizing polar baseband information is known as a polar modulator or polar transmitter. In this scheme, typically, the phase signal is fed to a DAC which controls the input of a PLL. This PLL is then used as a phase–frequency modulator. The output of this PLL is fed into a highly nonlinear but very efficient power amplifier. The amplitude information which is obtained through a different DAC adds the

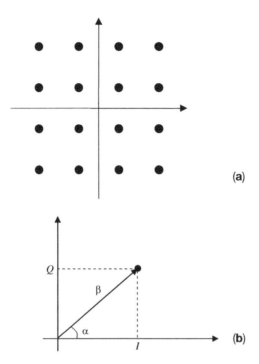

Figure 2.9 (**a**) General QAM-16 constellation diagram. (**b**) Representation of any of the points in the constellation diagram using Cartesian or polar coordinates.

amplitude modulation to the power amplifier.[38] The biggest advantage of a polar modulator is that it utilizes a highly efficient PA.

Polar modulators are the subject of very active research. In general, utilizing polar modulators for wide-bandwidth signals as well as signals that have a large PAR is quite challenging. WLAN OFDM signals are both very wide bandwidth and also possess a very large PAR. Therefore, at the date of publication, no commercial WLAN ICs utilize polar modulation. The interested reader can refer to the technical literature for more details on polar modulators.

2.2 PROCESS CHOICES: CMOS VERSUS SiGe BiCMOS

Let us briefly discuss the choice of process technology with a particular emphasis on the choice between CMOS and silicon germanium (SiGe)

[38]Various schemes can be utilized for this purpose. For example, the supply voltage of the PA can be modulated using a high efficiency DC–DC converter.

BiCMOS. There have been numerous discussions, publications, conference panels, and so on, on this topic with little uniform consensus. Further, although there may seem to be consensus on certain aspects at a given time, at a different time, based on the performance and cost of the two processes, a different popular opinion may emerge.

Despite the statements in the previous paragraph, certain factual statements can be made:

- Bipolar (SiGe) transistors are better for most (but not all) RF applications.
   1. They offer a higher transconductance for a given DC (g_m/I),[39] lower flicker noise ($1/f$ noise), and a higher breakdown voltages times f_T.[40]
   2. Bipolar SiGe devices typically offer better modeling and therefore often more reliable first-cut designs and faster time to market. There are two fundamental reasons for this. First, short-channel MOS devices are inherently much more complicated devices than the state-of-the-art bipolar devices and therefore often require nonphysical, numerically fit models to describe their operation. Second, quite often foundries offering the state-of-the-art CMOS processes target their models to digital designers and therefore do not spend as much resources or attention on the characteristics of their devices that apply to analog and RF design (these include characteristics such as flicker noise, thermal noise, Monte Carlo models, and nonlinearities). The SiGe foundries often directly cater to the analog/RF designer, and therefore not only spend more time modeling their devices but often design their devices for the particular needs of the analog/RF designer.
- The CMOS process does offer some advantages:
   1. CMOS offers a lower g_m/I than the bipolar process. This characteristic is useful for applications such as current sources where the de-

---

[39] A high g_m/I is often needed in analog/RF design where high gain and low power consumption are desired.

[40] F_t is the frequency at which the current gain of the transistor in the common emitter (or common source) configuration with the collector (or drain) AC grounded drops to unity. In other words, the maximum frequency at which the transistor can provide current gain. The high breakdown voltage multiplied by f_T (or f_{max}) is a very important device characteristic for power amplifier applications. Although the CMOS scaling is allowing CMOS devices to achieve extremely high f_T, at the same time the maximum operating voltage of these devices is being reduced due to issues with device stress and breakdown. Power amplifiers typically require simultaneous high breakdown voltage and high f_T, that is a characteristic at which SiGe and especially GaAs processes excel.

vice can amplify its own noise and degrade SNR of the system. The low g_m/I may be useful in VCOs also. The reason is that typically a large tail current source is needed in order to achieve a large swing and therefore low phase noise. However, the noise added by the cross-coupled devices near the zero crossing point is proportional to their transconductance and therefore a large signal g_m (G_m) "just large enough" to ensure oscillation is desired. There are examples of other particular applications in which a lower g_m/I is desirable.

2. It can be argued that CMOS devices have higher linearity than bipolar devices, especially in the velocity saturation region.[41] However, it can also be argued that since the bipolar devices have a higher f_T and g_m for a given DC, the excess gain and bandwidth can be used along with feedback techniques in order to linearize the device. So once again there is no simple answer.

3. CMOS devices offer a higher input impedance at low frequencies. This is quite useful for many baseband and low frequency circuits and also for biasing high frequency circuits. However, for a similar application, typically the total input capacitance (the capacitance seen looking into the gate) of a CMOS device would be larger than the total equivalent input capacitance (the capacitance seen looking into the base) of a bipolar device. Therefore at higher frequencies, often the input impedance of the CMOS device would be *lower* than the equivalent bipolar device.

4. The triode operation of the MOS device can be quite useful in certain applications. The MOS device can replace an otherwise very large physical (poly) resistor in a fraction of the area. Further a MOS device can be used as a variable resistor in certain applications.

5. MOS devices make good voltage switches and can be switched quite fast. The low frequency high impedance of the MOS devices and their switching capability make them ideal for mixed-signal and switched-capacitor circuits (e.g., ADCs, switch-cap filters). Bipolar devices make for poor voltage switches as, once forced into saturation, their reverse recovery time is quite long.

- Highly integrated transceivers today have significant digital content. It is clear, therefore, that there is a need for CMOS devices in such transceivers.

---

[41]Bipolar devices inherently have an exponential I/V characteristic and are therefore inherently quite nonlinear. CMOS devices with long channel lengths have a second-order I/V characteristic. Short-channel CMOS devices which normally operate in the velocity saturation region approach a linear I/V characteristic. However, often other sources of nonlinearity associated with short-channel devices can kick in and reduce the linearity of the device.

- One may conclude from the foregoing statements that BiCMOS should then be the process of choice. However:

  1. A BiCMOS fab is not as readily available, especially for fabless companies (although this is slowly changing)

  2. There is a cost premium associated with a BiCMOS process. It may be argued that the cost of a SiGe bipolar wafer (say with a 0.35μ CMOS option) is no more than that of a CMOS wafer (say 0.18μ straight CMOS process). The issue here is that one is paying a penalty in the performance of the digital devices and therefore it is not a fair comparison. In fact, the state-of-the-art CMOS process is often two or more generations ahead of the state-of-the-art SiGe BiCMOS process in terms of the CMOS devices. Given that digital applications crave for the best CMOS transistors, they often opt for the best (reasonably priced) digital process to help improve their performance and/or reduce their power consumption.

  3. At various times, it has been argued that CMOS will be taking over the world and wiping BiCMOS out.

  4. For certain standardized applications it may not be worth the extra premium to pay for BiCMOS, and simply meeting the standard *may* be good enough. In reality, whether or not this statement is true would depend on the market segment one is addressing. Certainly for some consumer-based markets, especially if power consumption is not a primary concern and if one is addressing the bottom and cost-sensitive part of the market, it would make sense to meet the bare minimum requirements and reduce the cost to the extent possible. However, if one is, for example, targeting the enterprise market where very high performance is expected and mediocrity is not tolerated, it may be necessary to obtain the highest performance possible. In such cases reducing cost would be a secondary objective.

For desktop and laptop computing needs, several leading suppliers have shown single-chip "vanilla" CMOS WLAN solutions with essentially all transceiver components integrated other than possibly the high power amplifiers. Interestingly, for the much more power-sensitive embedded market (e.g., WIFI for cellular phones), two trends are emerging. Certain vendors are perusing single-chip radio + MAC + PHY all-CMOS solutions and using an external SiGe or GaAs power amplifier and integrating the two chips in one package. Others are partitioning their solution differently: single-chip CMOS PHY+MAC and a separate SiGe BiCMOS PA + radio. The two chips are then integrated in one package. Like any other engineering problem, each approach has its merits and disadvantages.

In summary, it is generally accepted that for high frequency applications, in terms of pure performance, CMOS is not the process of choice. However, for many applications, with the help of a digital signal processor (DSP), many of the shortcomings of the CMOS process can be corrected for. As a result, for many applications a "high performance" and high yield CMOS solution is achievable and comes at a lower cost than its SiGe BiCMOS counterpart.

CHAPTER 3

Analog Impairments and Issues

In this chapter we will consider various analog impairments and physical constraints present in real radio systems and evaluate their impact on the system performance.[42] We will also evaluate the interrelation between these impairments, the choice of the radio architecture, and the choice of the process technology.

We will start by examining the receiver impairments. We will then transition and discuss the transmitter impairments. We will finally consider and discuss impairments relating to the VCO/LOGEN/PLL circuitry.

3.1 RECEIVER SENSITIVITY AND NOISE FIGURE

At the circuit level, one of the most important characteristics of a receiver is its noise figure. There are several definitions for the noise figure, and it can be easily shown that these definitions are equivalent. The noise figure can be defined as

$$NF = 10 \log \left(\frac{SNR_{in}}{SNR_{out}} \right)$$

Since the SNR at the output can never be larger than the SNR at the input (real circuits can only add noise; they cannot take away noise!), the minimum NF achievable is 0 dB (ratio of 1). This can only occur if the circuit under test adds absolutely no noise to the signal. In practice, this can never happen and NF is always a number greater than 0 dB.

From a system perspective, the receiver NF is directly related to the system sensitivity.[43] The lower the NF of a system, the lower (i.e., better) its sensitivity will be. Quite often, the standards do not specify the required NF

---

[42]Some system designers affectionately refer to the radio block as the "impairment generator" block!

[43]Sensitivity of a system is defined as the signal power level at which it can receive the required modulated signal at an error rate (bit error rate or packet error rate) determined by the standard.

Wireless LAN Radios: System Definition to Transistor Design. By Arya Behzad
Copyright © 2008 the Institute of Electrical and Electronics Engineers, Inc.

of the receiver. Instead, they specify the minimum required sensitivity for the various data rates allowed in the standard. In order to understand the relationship between the receiver NF and the system sensitivity, it is best to go through an example.

The 802.11a/g requires the sensitivity of the system to be better than –83 dBm for the 6-Mbps rate and better than –65 dBm for the 54-Mbps rate. At the same time, a typical (realistic) OFDM system requires approximately 20 dB of SNR[44] in order to be able to generate less than 10% PER (packet error rate, the limit required by the 802.11 standard to set the sensitivity levels). The equivalent required SNR for the 6 Mbps is approximately 5 dB.

With the information on minimum required SNR and the required sensitivity at hand, one could calculate the required NF of the system (in decibels):

$$\text{NF} = \text{sensitivity} - kT - 10 \log(\text{BW}) - \text{SNR} \qquad (3.1)$$

where k is the Boltzmann constant and T is the absolute temperature in kelvin. Therefore kT represents the thermal noise floor, which is approximately equal to –174 dBm/Hz at room temperature. Here, BW is the bandwidth of the modulated signal[45] and SNR is the required signal-to-noise ratio required at the input of the PHY slicer.

For a 6-Mbps data rate in a 802.11a/g system, the required SNR is 5 dB and the channel bandwidth is 16.3 MHz. Therefore,

$$\text{NF} = -83 - (-174) - 10 \log(16.3 \times 10^6) - 5$$

$$= 13.8 \text{ dB}$$

Of course note that this is the overall required NF for the entire system, including any board losses (a cascade NF equation can show that these losses directly add to the NF of the receiver). In a typical 5-GHz band, the front end losses can be 2 to 3 dB. Therefore the NF of the receiver would have to be in the range of 11 dB. Today, this is a fairly simple number to achieve. State-of-the-art CMOS radios in production today offer a NF in the range of 3.5 to 5.5 dB and a chip-referred sensitivity of approximately –93 dBm.

---

[44]It can be shown that the theoretical minimum number required for the 54-Mbps OFDM rate is 18 dB. In practice, various imperfections cause the minimum required SNR to be larger than 18 dB. Also note that this minimum SNR assumes a soft Viterbi decoder. A hard Viterbi decoding scheme would require 2 to 3 dB more SNR.

[45]In reality BW should be the bandwidth of the narrowest filter in the receiver chain before any effect that can cause noise folding. In a well-designed system, often this bandwidth is selected to be equal to the bandwidth of the modulated signal.

This is an example of a case in which simply meeting the standard is not good enough. A product that barely meets the standard would not be competitive in the market.

Note that it would not be a fair comparison to compare the sensitivity numbers of standards with different modulation bandwidths. As evident from Equation 3.1, reducing the channel bandwidth has a very significant impact on the sensitivity of the system when all other factors are kept constant. For example, in the cellular GSM, systems are achieving sensitivity levels well below –100 dBm. However, this is primarily because the modulation bandwidth is 100 kHz. This would translate into an "inherent" sensitivity advantage of 10 $\log(16.3 \times 10^6)$ – 10 $\log(100 \times 10^5)$, or 22 dB, as compared to the 802.11a/g standard (all else being the same). Further some proprietary 802.11 implementations use the narrow-banding technique in order to offer better sensitivity. Again, there is no magic to this approach. The radio performance has not changed. These solutions simply trade off data rate (and throughput) for range.

3.2   RECEIVER DC OFFSETS AND LO LEAKAGE

Another set of receiver-related impairments are issues related to DC offsets and LO leakage. In some aspects these impairments are related, since the leakage of the LO into the input port of the mixer can cause DC offsets, while this same leakage into the antenna port can cause LO reradiation.

Relatively small values of DC offsets can cause corruption of the baseband-modulated signal, while large values of DC offsets can cause the saturation of the following stages.

On the other hand, LO leakage back to the antenna would cause reradiation of the LO, which may violate the standard's requirements or FCC requirements. In a direct-conversion receiver, this LO would typically be at the center of the receiver desired band. In a low IF or superheterodyne receiver, the LO would typically fall on an adjacent band or an entirely different band. In either case, these radiations would be governed by the standard or the FCC since they would cause interference with other users.

Several ways by which DC offsets or reradiation can be created are shown in Figure 3.1. Coupling through path 1 would cause the LO to leak to the RF port of the mixer and can therefore create a DC offset. Coupling through path 2 is even more problematic than through path 1 since the LO at the RF port of the mixer is now amplified by the gain of the LNA and would therefore create DC offsets with larger amplitudes. A large received interfering signal as shown in path 3 can couple to the LO port and mix with it-

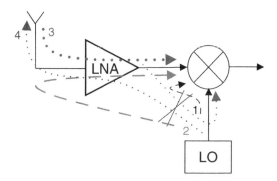

Figure 3.1   Simplified receiver front end depicting various LO paths that may result in generation of DC offsets at mixer output. For example, one source of signal corruption is due to large interferers self-mixing at the mixer (path 3) creating undesirable baseband components. This is much more of a problem for direct-conversion receivers. The path for LO reradiation through the antenna is also shown.

self and therefore also create DC offsets. The DC offset due to path 3 can be time variant because it is a function of the large incoming interferer. Finally, path 4 shows the coupling path that would cause LO reradiation. Note that the reverse isolation of the LNA would typically help reduce the LO reradiation problem, unless the coupling path is through some other path (e.g., through the package parasitics). It is also important to note that the LO leakage radiated through the antenna can reflect off of moving objects and be received at the antenna and cause a potentially strong time-varying DC offset.

Note that there are other mechanisms that can create DC offsets, including the even-order nonlinearities of the mixer, which will be discussed in more detail later.

These impairments and their impact on the system performance relate very strongly to the choice of radio architecture. First, in a direct-conversion receiver the LO is operating at a higher frequency than in a superheterodyne,[46] resulting in lower mixer port-to-port isolation. Further all parasitic coupling paths are stronger at higher frequencies. Therefore, with all else being equal, the amount of capacitive couplings would be larger in a direct-conversion receiver resulting in larger DC offsets. Further, all of the above mentioned impairments that result in DC offsets can impact a direct-conversion receiver quite severely. Recall that due to lack of significant amount of

---

[46]A "low injecting" LO superheterodyne architecture is assumed (i.e., one at which the LO frequency is lower than the RF frequency). A low injecting superheterodyne architecture is the most commonly used one. A high injecting superheterodyne architecture would have larger parasitic couplings at the RF sections than a direct-conversion architecture.

RF gain in a direct-conversion architecture, the baseband chain is required to possess a significant amount of gain. Therefore any DC offsets present at the output of the mixer (whether created due to LO leakages, nonlinearities, or device mismatches) would be amplified by this large gain and would likely saturate the following stages. Further the DC offsets lie in the center of the desired modulated signal and can corrupt this desired signal. It is often required, therefore, that these DC offsets be removed. Several schemes have been proposed for removing these DC offsets. These typically include some form of high pass filtering or the injection of a correction DC signal in the signal path. Most schemes cancel most or all of the DC offsets in the analog domain. Others rely on digital signal processing for removing some or all of the DC offsets. If a digital scheme is used, it is important to ensure that the analog baseband signal path will not be saturated by the DC offsets and that the ADC has sufficient excess dynamic range for large DC offsets. For most practical WLAN direct-conversion receivers, the DC offsets have to be canceled in the analog domain. This is because of a very high amount of gain in the receive signal path and the potential for clipping due to DC offsets.

As described above, not only the amount of DC offsets generated in a superheterodyne receiver would be smaller than those of a direct-conversion receiver, but also their impact on the system performance is typically insignificant due to the lower gain of the receive chain at the baseband frequencies.

Similar factors limit the LO reradiation in a direct-conversion receiver. The fact that the LO operates at a higher frequency that the superheterodyne receiver[47] causes larger coupling of the LO to the antenna port and therefore a larger LO reradiation. However, the fact that the LO reradiation is in the center of the desired channel may translate into looser specifications imposed by the standard or the FCC. If a band-select filter is not used between the antenna and the LNA, and depending on the choice of intermediate frequency, a superheterodyne receiver may have the most severe LO reradiation specifications. However, if a band-select filter is utilized, the reradiated LO signal would be significantly attenuated by this filter. With a band-select filter, a low IF receiver would probably impose the most stringent requirements on the LO reradiation specifications (since the frequency of the LO is quite high and the limited Q band-select filter cannot attenuate the LO signal).

For the most part the same remedies can be applied to solving the DC-offset-related issues as well as the LO reradiation issues.

---

[47]Again a low side-injecting superheterodyne architecture is assumed.

Good matching and layout are critical to reducing asymmetries than can translate into DC offset generation and differential leakages. However, even with the best reasonable layouts, for a direct-conversion receiver, some form of DC offset compensation should be used.

Two popular approaches exist for DC offset cancellation at baseband. The first approach relies on sampling the offsets at low gain settings (such that the receive chain is not saturated) and compensating the offsets at the proper points through the receive chain with high resolution DACs.[48] In this scheme, the system relies on the high absolute accuracy of the baseband gain settings to calculate and apply the proper DC offset cancellation value as the baseband gain settings change during the receiver operation. If the DC offsets are a function of temperature, the DC correction process has to be repeated periodically or by monitoring a temperature sensor. The advantage of this scheme is that in normal operation there are no actual high pass filters in the receive chain and therefore the EVM of the inner OFDM subcarriers is not adversely impacted. The disadvantage of this technique is that it is an open-loop scheme which relies on certain assumptions. As long as it can be guaranteed that the assumptions would hold under all operating conditions, this scheme works well.

An alternative approach to solve the baseband DC offset problems in a direct-conversion receiver is to use some form of a HPF. The HPFs can be implemented in a variety of ways, including simple AC coupling and DC offset cancellation loop. In either case, the proper placement of the location of the HPF pole is critical. The filter poles cannot be placed too high in frequency, since this would result in the attenuation of the low index subcarriers. If the HPF pole is placed too high in frequency, it may cause a severe degradation in the EVM of the lower subcarriers due to significant group delay variation at the lower OFDM subcarriers. At the same time, the filter poles cannot be placed too low in frequency, because this would result in very slow transient response. Given the limited amount of time present in the 802.11 preambles and the many actions that need to be taken during this time, it is important that any transients in the system settle quickly. Since no single pole location may be able to settle both these requirements simultaneously, an "agile" HPF pole may be required. Such an agile pole would jump to a high frequency during preamble (the preamble is quite robust to impairments) in order to allow for fast settling and then jump to a low frequency during the payload in order to preserve the payload content. The advantage of the HPF method of DC rejection is that it relies on a DC offset cancella-

---

[48]Alternatively offset values can be estimated at multiple gain settings for higher accuracy offset cancellation at the expense of longer calibration time.

tion scheme that is continuously tracking DC offsets and canceling them. On the other hand, the HPFs will impact the quality of the inner OFDM subcarriers and can cause degradation in overall performance. The degree of the EVM degradation would depend on many factors, including the number of HPF poles in the baseband section, the location of the poles, and whether agile HPFs are used.

If HPFs are used for DC cancellation purposes in the receiver baseband section, the optimal system performance (especially in the presence of frequency offsets) may require the use of a mixed-mode automatic frequency control loop. More will be discussed on this topic in Chapters 6 and 7.

In order to improve the LO reradiation issue in a direct-conversion receiver, in addition to the good layout techniques mentioned, the use of a high reverse isolation LNA would be useful (assuming the coupling is through the LNA).

3.3 RECEIVER FLICKER NOISE

Another receiver impairment is the flicker, or $1/f$, noise. Flicker noise is a form of noise which, like the other forms of noise, limits the SNR of the receiver. However, unlike thermal noise, flicker noise has a low pass characteristic, as the example in Figure 3.2b depicts. Assuming a flat frequency response for the received desired channel, the SNR of the received signal would have a high pass characteristic. For an OFDM-modulated signal (with flat frequency response) subjected to strong flicker noise, this would translate into the lower indexed subcarriers having a lower SNR than the higher indexed subcarriers.

Figure 3.2b shows an example of the characteristics of flicker noise typical of a CMOS process. In this example, the intersection of the point where the thermal noise floor crosses the flicker noise line is called the *flicker noise corner*. For a "typical" CMOS process this corner frequency can be around 100 kHz (as shown in this example). However, this corner is a function of several major factors which can easily change the corner frequency by an order of magnitude. These factors include the thermal noise floor (the lower the thermal noise floor, the *higher* the flicker noise corner frequency) and the amount of flicker noise (the higher the amount of flicker noise, the higher the flicker noise corner frequency).

It is also important to note that, although another name for flicker noise is $1/f$ noise, flicker noise does not always have a $1/f$ frequency dependence. The value for the flicker noise exponent would determine the slope of the flicker noise as a function of frequency. Typical values for this exponent

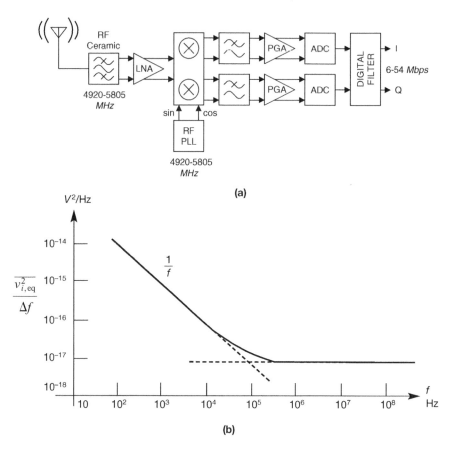

Figure 3.2 (**a**) Direct-conversion receiver architecture block diagram showing relatively low amount of gain before hitting high flicker noise baseband section. (**b**) example of flicker noise characteristics. (Figure 3.2b after Gray and Meyer, *Analysis and Design of Analog Integrated Circuits,* Wiley, New York, 1994.)

range from 0.5 to 1.5. A general relation for a MOS field-effect transistor (MOSFET) flicker noise voltage referred to the gate of the device is given by

$$\overline{v_{i,\text{eq}}^2} = \frac{K\,\overline{\Delta f}}{C_{\text{ox}}WLf^n}$$

where K is the flicker noise coefficient, C_{ox} is the gate oxide unit capacitance, W is the width of the device, L is the length of the device, f is the frequency of interest, and n is the frequency exponent (often equal to or close to unity). Note that this is a very simple relation that shows the behavior of the device flicker noise as a function of device and process characteristics.

However, flicker noise in a MOSFET device is a very complicated phenomenon and is an area of active research. For example, it is generally accepted that the dominant mechanisms causing flicker noise in NMOS devices are surface defects and scattering phenomena (the so-called number fluctuation theory). However, the dominant mechanisms causing flicker noise in a PMOS device are bulk crystal defects (the so-called mobility fluctuation theory).[49] So it is not unexpected that PMOS devices have a significantly different flicker noise magnitude and characteristic than NMOS devices. As it turns out, for modern CMOS processes, PMOS devices typically have much lower flicker noise than their NMOS counterparts. It is interesting to also note that "native" devices[50] often have lower flicker noise characteristics than a similarly sized normal device in the same process.

Also note that the foregoing equation indicated that the input referred flicker noise of a MOSFET device is bias independent. In reality, especially for short-channel MOSFET devices, this is not the case. Also, note that the exponent n may have a value other than unity.

It is clear from the foregoing that the designer has many knobs to adjust in designing a CMOS circuit in order to improve the flicker noise performance of the circuit. However, often trade-offs have to be made (size, power consumption, bandwidth of circuit, etc.) which limit the capability of the designer in reducing the impact of flicker noise. Furthermore fairly high flicker noise levels are part of the practical characteristics of today's modern CMOS processes.

In general the above equation gives a first-order indication of the parameters that are available to the designer in order to reduce the flicker noise of a device, but for more accurate estimates more sophisticated equations and device modeling need to be used. The more advanced CMOS device models such as BSIM4 have many more parameters in order to obtain a better fit for the flicker noise of the MOSFET devices.

In contrast to CMOS devices, bipolar devices often have much lower flicker noise corners. Modern bipolar processes exhibit flicker noise corners in the range of 10 to 1000 Hz and are therefore much preferred to CMOS devices from a flicker noise point of view. One possible explanation for the low flicker noise corner of bipolar devices as compared to

---

[49]This difference between NMOS and PMOS devices seems to make sense given that it is generally believed that NMOS devices are surface channel devices where the carriers have more interaction with the $Si–SiO_2$ surface states, while PMOS devices are buried channel devices interacting primarily with the lattice. Typically carrier operation below the surface results in much less flicker noise.

[50]Native devices are devices that have not been subjected to the threshold adjust implant and therefore have the native threshold voltage of the process. This native threshold voltage is often close to 0 V, making the device behave almost like a depletion device.

CMOS devices is that state-of-the-art bipolar devices are inherently vertical devices (i.e., the transistor action occurs in the vertical dimension: see cross section in Figure 3.3a), whereas MOS devices, especially NMOS devices, are much more of surface devices (i.e., the transistor action occurs on the surface of the device underneath the gate: see Fig. 3.3b). Since during the processing of the wafers often processing debris "floats" to the top and gets trapped at the surface, this causes many more surface defects in the MOS device processing. These defects in turn give rise to higher flicker noise.

How does the flicker noise impact the choice of the receiver architecture? In a superheterodyne receiver, due to the high amount of total gain at the RF and IF stages, the signal is well amplified before encountering the high flicker noise blocks of the baseband circuits. So, in general, flicker noise is not a big issue for a superheterodyne architecture. In contrast, in a direct-conversion receiver such as that of Figure 3.2a, a minimum amount of gain is available at the RF front end. As such the flicker noise of the baseband blocks (mixer switching devices, filters, PGAs, ADCs, etc.), especially in a

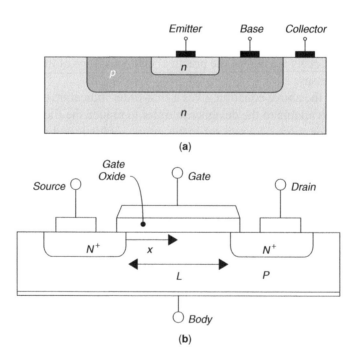

Figure 3.3 (a) Simplified cross section of bipolar junction transistor. Primary "transistor action" occurs in the vertical dimension. (b) Simplified cross section of N-MOSFET. Primary action occurs in the horizontal dimension.

CMOS implementation, can add significantly to the noise level of the low frequency content of the modulated desired signal.

Another factor that determines the extent of the impact of the flicker noise on the received signal quality in a direct-conversion receiver is the bandwidth of the desired channel. For a desired signal with a wide bandwidth, the NF (and therefore SNR) would degrade for the lower frequency content, but averaged over the wideband, the signal quality may not be significantly impacted. For a narrowband signal, on the other hand, the entire bandwidth of the signal may be in the flicker-noise-dominated region of the spectrum, and an appreciable impact on the average NF of the system will be observed.

It is instructive to take a numerical example for a couple of standards. First consider the 802.11a signal. When down converted to baseband in a typical direct-conversion receiver, the highest frequency content of the desired signal is at 8.125 MHz. With a typical flicker noise corner of, say, 400 kHz, only the lowest two OFDM subcarriers will be impacted by this flicker noise, and on average only a marginal hit to the NF will result. On the other hand, consider the same receiver receiving a GSM/EDGE-modulated signal, whose highest frequency content after down conversion to baseband is 100 kHz (since the RF signal bandwidth in GSM/EDGE is 200 kHz). Clearly the average NF of this system is quite severely impacted by the flicker noise. This is one fundamental reason (among others) why most GSM receivers use a low IF architecture with the IF at about 150 kHz rather than a direct-conversion architecture.

3.4 RECEIVER INTERFERERS AND INTERMODULATION DISTORTION

Another set of receiver impairments is related to the nonlinearities of the receiver. An ideal receiver would faithfully amplify, frequency down convert, and filter the desired signal without creating any nonlinear components at the output. In reality, nonlinearities do exist in a practical receiver and as a result distortion components are generated. The impact of these nonlinear components is that they impact the EVM (and therefore quality) of the received signal and can increase the PERs in the receiver.

3.4.1 IP3, IP2, and P1dB

Parameters often used to characterize the nonlinearity of a circuit are the harmonic distortion (HD) and intermodulation distortion (IM) numbers. Specifically, the second- and third-order harmonic distortion numbers (HD2 and HD3) and intermodulation distortion numbers (IM2 and IM3) are of great interest in characterizing circuits operating under nonlinear conditions.

It can easily be shown that, by passing a single-tone sine wave signal with frequency f_1 through a circuit with second-order nonlinearities, a component with twice the frequency at $2f_1$ is created. This component is referred to as the HD2 and it is usually specified in terms of its amplitude relative to the desired signal in decibels relative to the carrier (dBc). Similarly, by passing *two* equal-amplitude sine wave signals with frequencies f_1 and f_2 through the same circuit, frequency components at $f_1 - f_2$ and $f_2 - f_1$ would be created.[51] These components are referred to as IM2 components. IM2 components are also typically specified in terms of their amplitude relative to each of the desired tones (in dBc). Note that even-order nonlinear terms can also generate DC terms. Figure 3.4 describes these concepts pictorially. Note that second-order nonlinear components (HD2 and IM2) in a non-frequency-translating and non-baseband circuit can be rejected by using a bandpass filter and *may* therefore not cause significant problems at the system level. It is also important to note that even-order nonlinearities (including HD2 and IM2) in a circuit can be minimized by using differential circuits. Ideal differential circuits will have infinite rejection of even-order harmonics since they possess an odd transfer function. Mathematically it can be shown that the power series expansion of an odd transfer function will have no even-order terms. Of course, in a practical circuit, imperfections in the layout symmetry would cause finite levels of even-order distortion components.

Similarly, by passing a single tone at frequency f_1 through a circuit with third-order nonlinearities, components at $3f_1$ are generated (HD3). By passing two tones with frequencies at f_1 and f_2, nonlinear components (IM3) at $2f_1 - f_2$ and $2f_2 - f_1$ are generated. It is important that, although HD3 components can typically be rejected by filtering, IM3 components can fall in band (or very close in to the band) and are therefore difficult to practically filter. Close attention, therefore, should be given to reducing the amount of third-order nonlinearities of the circuit.

Parameters often used to characterize the nonlinearities of the receiver are the second-order intercept point (IP2), the third-order intercept point (IP3), and in general the nth-order intercept point (IPn). The linearity of an amplifier is easily characterized by its intercept point. The second-order intercept point quantifies second-order linearity performance while the third-order intercept point quantifies third-order linearity. Figure 3.4 displays a

---

[51]We are assuming non-frequency-translating circuits at this point. We will discuss frequency-translating circuits (like mixers) later.

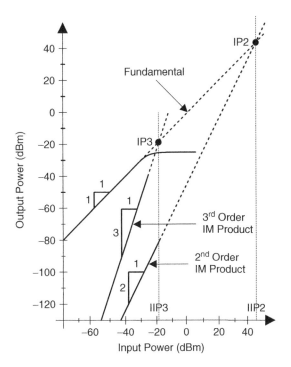

Figure 3.4 Extrapolation of linear term, second-order IM product, and third-order IM product to obtain IP2 and IP3 points. Given the IP2 and IP3 terms and the input power, the amount of intermodulation distortion in dBc can be calculated for "simple" nonlinearities.

typical plot of the desired signal (fundamental) as well as the second- and third-order intermodulation products as the input power is swept on the x axis. The fundamental will increase at the same rate as the input power. The second-order intermodulation product will increase 2 dB for every 1-dB increase in the input power and the third-order intermodulation product will increase by 3 dB for every 1-dB increase in input power.[52] The nth intercept point is the point at which the extrapolated nth-order intermodulation product intersects the fundamental as the input power is swept. The IP2 and IP3 points are labeled in Figure 3.4 according to this definition. Notice that, since the slopes of the curves are known, the intercept points can be calculated from a single point without actually performing the swept measurement. This ease of measurement is why intercept points are so valuable as a way to quantify performance.

---

[52]This is due to the fact that the second-order distortion is due to an P_{in}^2 term in the power series expansion of the transfer function. An P_{in}^2 term, in the log domain (dB) would have a 2-dB/dB slope. The same explanation would apply to the nth-order distortion terms.

A few important points need to be made at this point. Most comments below are made in reference to IP2 and/or IP3, but the general concepts are applicable to IPn:

- IP2 and IP3 extrapolations as described above only apply for circuits with "simple" nonlinearities that are characterized by a power series expansion of the transfer function. These extrapolations no longer apply if the transfer function includes complex nonlinearities such as kinks, crossover distortion, and hard clipping. For example, if the transfer function includes a change in the region of operation as a function of the input power, the theoretical IP2 and IP3 concepts as described do not apply for large signals (they may still apply over a limited region of the transfer function).
- The linear gain term as well as the third-order distortion term "compresses" well before the actual IP3 point (this is shown for the linear term but not for the third-order term in Fig. 3.4). In other words, the IP3 point is only a theoretical extrapolated point. In a real circuit this point is never achieved. For example, in Figure 3.4, if an input power of –20 dBm is applied, an output linear power of –20 dBm will *not* be observed. Further a third-order IM term of –20 dBm will not be observed. IP3 is only an *extrapolated* value based on small input levels where weak nonlinearities dominate the circuit's behavior. At large input levels, where the input level is close to the IP3 value, strong nonlinearities dominate (quite often dominated by higher order nonlinearities such as IM5 and IM7).
- The input power level at which the IP3 point is reached is called the input IP3 or IIP3. The corresponding output power level is known as the output IP3 or OIP3. The relation between the IIP3 and OIP3 is given by the following simple equation: OIP3 = gain + IIP3.
- For receiver subblocks, the linearity is often expressed in terms of input-related quantities (in this case IIP3). This is because the IIP3 is an indication of the largest input signal the receiver can receive before running into issues related to nonlinearities. On the other hand, for transmitter subblocks, the linearity if often expressed in terms of output-related quantities (in this case OIP3). This is because in the case of a transmitter the quantity of interest is the largest power the transmitter can emit before running into issues based on nonlinearities.
- Using basic geometric relations and Figure 3.4 it can be shown that the input and output IP3 follow the following relations:

$$\text{IIP3} = P_{\text{in}} + \tfrac{1}{2}(\Delta\text{IM3}) \qquad \text{OIP3} = P_{\text{out}} + \tfrac{1}{2}(\Delta\text{IM3}) \tag{3.2}$$

Similarly it can be shown that for IP2

$$\text{IIP2} = P_{\text{in}} + (\Delta\text{IM2}) \qquad \text{OIP2} = P_{\text{out}} + (\Delta\text{IM2}) \qquad (3.3)$$

These relations can be used to calculate the relevant intercept points given a single measured (or simulated) IM value and the input or output power level. Equivalently, given the intercept points and the power level, the relevant IM value can easily be calculated.

- In a frequency-translating receiver circuit (a receiver that includes a mixer in the lineup), the IM3 components that are of interest would fall at the following frequencies: $2f_1 - f_2 - f_{\text{LO}}$ and $2f_2 - f_1 - f_{\text{LO}}$.
- In a frequency-translating receiver circuit (a receiver that includes a mixer in the lineup), the IM2 components that are of interest would fall at the following frequencies: $f_1 - f_2$ and $f_2 - f_1$.[53] It is interesting to note that these components would only be of interest in a direct-conversion receiver and can easily be rejected by a filter in other receiver architectures. Even in a direct-conversion receiver, given a perfectly differential mixer (but a nonlinear front end), the even-order harmonics will be rejected. In reality, however, due to imperfections in the layout and the manufacturing process, a finite amount of IM2 components will be generated.

Most standards, including the 802.11 standard, do *not* specify any IP3 or IP2 (or IM3 or IM2) requirements. Instead they often specify two quantities:

1. The maximum level of *desired* signal the receiver needs to be able to handle.
2. The maximum level of *interference* signal the receiver is required to be able to handle while receiving a "weak" desired signal. Often, this requirement is specified in such a way that in the presence of an "interferer" the receiver should be able to perform at no worse than 3 dB above the sensitivity level (while no interference is present). The definition of an interferer varies from standard to standard.

In the case of 802.11a/g the receiver is required to be able to receive a maximum desired signal level (at the antenna) of at least −30 dBm for all the modulation types with a PER of no more than 10% for 1000-byte packets. Clearly, a typical practical receiver would provide a higher maximum re-

---

[53]An IM2 component at $f_1 + f_2 - 2f_{\text{LO}}$ can also be generated. Since the amplitude of $2f_{\text{LO}}$ is normally small, this IM2 component is typically negligible.

ceived signal capability for the lower data rates. In reality many of today's receivers are capable of performing significantly better than this by utilizing various gain control techniques and due to their high linearity levels. The equivalent maximum required capability of the receiver in the 802.11b standard is –10 dBm for a frame error rate (FER) of 8%.

802.11a/g specifies the receiver should be able to receive a desired signal at 3 dB above the sensitivity at the presence of an OFDM interferer 20 MHz (25 MHz for 802.11g) away at –63 dBm. This is known as the adjacent channel interferer (ACI) requirement. Further it requires the same 3-dB degraded sensitivity level in the presence of an OFDM interferer 40 MHz away at –47 dBm power level. This is known as the alternate adjacent channel interference (AACI) level requirement. These requirements apply to all OFDM rates. However, in the relative decibel scale, the requirement is more stringent for the lower data rates than it is for the higher data rates. As an example, a 6-Mbps desired signal is required to have +16 dB of ACI, whereas a 54-Mbps signal is only required to have –1 dB of ACI.

For the 802.11b standard, the interference is specified to have a power 35 dB above the desired signal at 25 MHz away while receiving a desired 11-Mbps CCK signal at 6 dB above the required sensitivity level (i.e., –70 dBm) at less than 8% FER.

How do these system level requirements relate to the IP2 and IP3 of the receiver? This is a fairly complex issue that deserves more discussion. The following important points need to be considered:

- Two-tone IP3 and IP2 values for a circuit are *indicative* of the interference tolerance of the system but certainly do not describe the whole picture. Similarly they provide information on how large of a (desired) modulated signal the system can tolerate but once again does not provide the whole picture.

- A two-tone CW signal has a PAR of 6 dB. This is far different that the PAR of an 802.11 OFDM-modulated signal which can have theoretical PARs of as much as 17 dB. Since nonlinearities present in a circuit typically respond to large instantaneous signals present in the input, it should not be surprising that they would respond quite differently to an OFDM signal that has the same average power as a two-tone CW signal.

- An 802.11 OFDM signal can be approximated by summing 52 CW signals in what is known as a multitone test. However, if the tones in a multitone are phase coherent, very large peak signals can be generated. Applying such a signal to the circuit would likely result in pessimistic predictions of the behavior of the circuit in the presence of a real OFDM-modulated signal. On the other hand, if a multitone test with random phases are used, the results may vary depending on the resultant phase coherence of the generated tones.

- An OFDM-modulated signal is a much more complex waveform than a two-tone CW signal. The statistical nature of the OFDM signal means that the peaks occur at a nondeterministic rate. This makes the comparison with a deterministic two-tone test even more complicated. The response of the circuit to an OFDM-modulated signal (and the resultant EVM) would depend not only on the peak signal but also on the frequency of these occurrences.

- Because of the general complications outlined above, the relation between a two-tone IP3 and the EVM of an 802.11 OFDM-modulated signal can not be expressed in a simple formula. Issues such as the types of nonlinearities in both the phase and the amplitude and the degree of soft versus hard limiting of the circuits impact the amount of backoff needed from the IP3 point of the circuit in order to successfully receive an OFDM-modulated signal. As a general rule of thumb 15 to 20 dB of backoff from the IP3 point of the circuit would be required in order to achieve a PER of less than 10% on a 54-Mbps OFDM-modulated signal. Certain digital postdistortion techniques may be applied to the signal in order to improve the tolerance of the system to nonlinearities experienced in the analog/RF sections of the receiver. It is also important to note that, as expected, the highest data rates with their high order modulations are much more sensitive to phase and amplitude nonlinearities. The lowest data rates with BPSK- and QPSK-type modulations are quite tolerant of nonlinearities, including severe amounts of clipping in the receiver.

We will now consider an example. Assume a two-tone –33-dBm/tone interferer at 50 MHz offset from the desired signal applied to the input of a LNA.[54] Further assume that the desired sensitivity level in the presence of the interferer is –78 dBm and that an SNDR of 10 dB is required on the de-

---

[54]Note that this is simply an example and not a real requirement set by the 802.11 standard. Further, note that for the sake of simplicity we will consider both the interferer and the desired signals as CW tones. In reality, for 802.11, the interference is specified as a modulated signal interfering with a desired modulated signal.

Different modulated signals will have different degrees of tolerance to narrowband versus broadband interferers. In general, OFDM signals are much more resilient to narrowband interferers. A narrowband interferer impacting an OFDM-modulated signal can be viewed as a multipath faded channel. Since the OFDM signal is designed to have tolerance against such a channel, it would naturally be resilient against such interference also.

sired signal. Finally assume that in this mode the system performance is limited purely by nonlinearities and that the noise floor is significantly below −88 dBm. Given Equation 3.1 we can calculate the overall required IIP3 for the LNA (Fig. 3.5a):

$$IIP3 = -33 + \tfrac{1}{2}(88 - 33)$$

$$= -5.5 \text{ dBm}$$

It is important to note that in a typical wireless receiver at maximum front-end gain settings the LNA is typically *not* the limiting factor in setting the linearity of the receiver. This topic will be discussed in more detail shortly.

Consider a second example. We will make the same assumptions as the previous example but apply the signal to the mixer of a direct-conversion receiver. As shown in Figure 3.5b, and explained earlier, the two tones and the mixer second-order nonlinearities would generate a component at baseband as shown in Figure 3.5b. Assuming that the applied tones are at frequencies f_1 and f_2, the IM2 component would fall at $f_1 - f_2$. The magnitude of this component referred to the input of the mixer would be

$$IIP2 = -33 + (88 - 33)$$

$$= +22 \text{ dBm}$$

Of course the calculated IIP2 and IIP3 numbers in the above examples are the minimum acceptable levels for these terms.

As a general note, some of the most stringent IIP2 requirements for a direct-conversion receiver are set by cellular standards such as GSM/EDGE/ WCDMA. This is because the maximum power of an interfering signal as well as the maximum ratio of the interferer power to the desired channel power can be quite large in the GSM/EDGE/WCDMA stan-

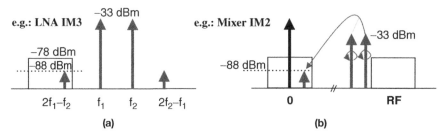

Figure 3.5 (a) IM3 example showing impact on LNA. (b) IM2 example showing impact on mixer in direct-conversion receiver.

dard. As compared to such systems, the IIP2 requirements for an 802.11 direct-conversion receiver are not very stringent.

Further, some of the most stringent IIP3 requirements for a receiver are set by full-duplex cellular standards such as WCDMA. In these systems, the receiver and the transmitter operate at the same time but on different (but close-by) frequency bands. A frequency duplexer is used to isolate the receiver from the high power levels of the transmitter. Even with this duplexer in place and 50 to 60 dB of isolation, the receiver is required to receive very small desired signals (<–100 dBm) in the presence of such a large "interferer." As compared to such systems, the IIP3 requirements for an 802.11 direct-conversion receiver are not very stringent.

An exception to the above noted examples in which the linearity of the WLAN receiver is quite important is in cases where the WLAN is to be operated along with other transceivers. For example, in a WLAN-enabled cell phone application, the WLAN receiver is to be able to operate despite the presence of large signal transmitters on the cellular radio. In the WLAN, for example, the transmitter may be transmitting a +26-dBm signal at 2.2 GHz, while the WLAN is trying to receive a –93-dBm signal at 2.4 GHz. Clearly a large degree of rejection of the cellular transmitter and a high linearity front-end WLAN receiver are required in such a case.

As will be apparent in the next few paragraphs, in many receiver systems, including those for 802.11, the typical linearity limiting block is the RF front end, in particular the RF mixer.

To make this clear, let us reexamine the lineup of a typical 802.11a direct-conversion receiver as shown in Figure 3.6.

The selectivity of the antenna, the matching network (not shown), and the optional bandpass filter attenuate the far-out interferers. At the same time, the baseband low pass filters reject the close-in interferers. The LNA and the mixer have to tolerate the interferers. Furthermore, as it will become apparent, due to the gain of the LNA, the RF mixer is typically the block that sets the final linearity performance limits in the receiver.

Above the block diagram in Figure 3.6, the typical voltage conversion gain figures are shown in the first row, with a "typical" LNA maximum gain of 25 dB and a typical active mixer gain of 6 dB. The second row displays typical IIP3 levels for the LNA and the mixer. Finally, the third row displays typical IIP2 levels for these blocks.[55] In this example, (IIP3mixer –

---

[55]Note that these numbers can vary significantly from design to design. The numbers used for this example are "typical" numbers based on fairly traditional designs reported in the literature. However, various (nontraditional) linearization techniques have been reported to improve the IIP3 of the mixer. Further, many calibration techniques have been used to improve the IIP2 of the mixer. With such techniques applied, the mixer may no longer be the limiting factor in the linearity performance of the receiver.

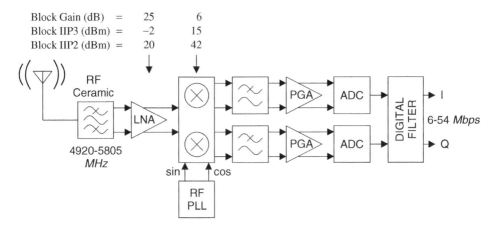

Figure 3.6 Block diagram of simplified 802.11a receiver with gain, IIP3, and IIP3 of LNA and mixers identified ("typical" numbers assumed). Given these numbers, it should be clear that in this system the mixer would limit the linearity of the receiver. Gain control in the LNA can be utilized to reduce the gain and reduce the impact of mixer linearity on system performance. Clearly, there would be a trade-off between NF and linearity in such a state.

IIP3lna) < Gain_lna. It should be clear that the mixer linearity would limit the overall IIP3 performance of the receiver. Similarly, (IIP2mixer − IIP2lna) < Gain_lna. The mixer in this example also limits the overall IIP2 performance of the receiver. Stated differently, the mixer in this example would *not* be a limiting factor in the linearity performance of the receiver *if* it would satisfy the following requirements: IIP3mixer ≫ +23 dBm and IIP2mixer ≫ +45 dBm. Without special design techniques, achieving these numbers, especially at the 5-GHz band, without sacrificing some other performance measure is difficult.

As shown in this example, the mixer is the limiting block in the linearity performance of the receiver when the LNA is at its higher gain settings. Under conditions in which the system can back off the LNA gain, the mixer may no longer be the limiting block in the linearity performance of the receiver. Under these conditions, the LNA will become the limiting block. However, at lower gain settings, it is typically easier to achieve higher IIP3 performance for the LNA. The toughest requirements on the receiver are set for the condition at which the receiver is required to receive signals at or close to the sensitivity level while being jammed by a large interferer. Under these conditions, simultaneous low NF and high linearity conditions are required to be met.

It is worth mentioning that front-end linearization techniques, in particular those applicable to linearization of the front-end mixer, are a very hot research area. A high linearity RF front end (hence mixer) is essential to the design of a receiver front end for a "software radio": a radio which can handle multiple standards by being programmed to the proper mode.

As in any other engineering problem, the designer faces trade-offs in designing the receiver for high linearity.

One or more of the following general trade-offs exist in trying to achieve a high IIP3:

- **High NF** Many IIP3 improvement techniques rely on resistive feedback techniques or the use of linearization devices. These devices contribute to the noise of the block, resulting in a higher NF.

- **High Power Consumption** In many cases, obtaining a high IIP3 requires operating the device at reasonably high current densities. For example, the linearity of a long-channel MOSFET device is related to the $V_{gt} (= V_{gs} - V_t)$ of the devices. In order to obtain a high V_{gt} for a given device size, one needs to burn more DC in the device.[56] Additionally, common open loop or closed loop linearization techniques require DC power of their own, increasing the overall power consumption of the block.[57]

- **Complexity** Many open- and closed-loop linearization techniques have been proposed in the literature. In general, these techniques will add complexity to the circuit, requiring more analysis and possibly delaying time to market.

Similarly, the following techniques help improve the IIP2 performance of the receiver. The trade-offs relating to the specified techniques are also listed below:

- **Differential Circuits** By utilizing differential circuits, the IIP2 of a circuit can be improved quite significantly. The trade-off is an increase of as much as 2 times in the area and power consumption.

- **Good Layout** A differential circuit is only as good as its layout! Careful attention to layout and matching details between the differential parts of the circuit can improve IP2. Further, in blocks such as mixers, attention to the layout of the switching devices would help reduce the leakage of the IIP2 terms to the baseband output. The trade-off to a good layout is the requirement for more resources and/or an increase in time to market.

---

[56]Note that for real MOSFETs, especially short-channel ones, excessively large quiescent current could actually *reduce* the IIP3 of the device due to effects such as the nonlinearity of the output impedance r_{ds}.

[57]Linearization techniques often offer a much better power efficiency than brute-force techniques such as increasing a V_{gt} of a MOSFET.

- **Large Devices** Using large devices would result in better matching between the differential portions of the circuit. The clear trade-offs include larger die area (cost) and higher power consumption.

- **Good Balun** A good on- or off-chip balun would ensure that the signal applied to the input ports of an LNA are fully differential with minimal mismatches on the balanced ports of the balun. This in turn results in better rejection of the IIP2 terms by the receiver chain. Trade-offs to this requirement are a higher cost and possibly larger size for the balun.

- **AC Coupling between LNA and Mixer** By using AC coupling between the LNA and the mixer, the IIP2 of the mixer block becomes essentially irrelevant. The AC-coupling capacitors reject any low frequency IM2 components generated by the LNA block, essentially making the mixer by far the limiting block in the IIP2 performance of a direct-conversion receiver.

Figure 3.7 displays a hypothetical case in a system with an overall IIP2 of +50 dBm. Note that this is a *very* high IIP2 number and that such a high per-

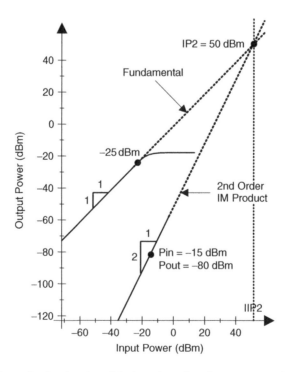

Figure 3.7 Example of estimation of IM2 product given input power and IP2. In this case two CW input interferers at –15 dBm each generate a baseband signal at –80 dBm.

formance typically requires the use of special design and layout techniques and/or some form of calibration. When two –15-dBm CW input signals are applied to this circuit, a –80-dBm IM2 tone is generated. Depending on the frequency of the two CW RF interferers and the magnitude of the desired RF signal, this can cause a significant degradation of the quality of the received signal.

Figure 3.8 displays the case of a receiver being exposed to a large cellular blocker signal while trying to receive a small desired OFDM WLAN signal. This scenario is common when operating a WLAN system in a cell phone application where the cellular band power amplifier can be transmitting in excess of +26 dBm while the WLAN receiver is in the receive mode. It is clear that in such cases a very large IIP2 for the WLAN receiver would be highly desirable in order to maintain the integrity of the received OFDM signal. Note that an amplitude-modulated cellular interferer would generate a copy of the interferer (with twice its bandwidth) at baseband, while a phase-modulated cellular interferer would cause a DC offset at baseband.

A related quantity to IIP3 used for measuring the linearity of a circuit is the 1-dB compression point, or P1dB. The pictorial definition of P1dB is shown in Figure 3.9. Based on this definition, the P1dB is the point at which the output power of the block is 1 dB less than what it would have been for an ideal (distortion-free) block. Stated differently, the P1dB is the point at which the gain of the block has dropped by 1 dB (due to compression) as compared to its small signal gain level. The 1-dB compression point of a circuit is closely related to odd-order nonlinearities in that circuit. For "simple nonlinearities" it can mathematically be shown that the P1dB of a circuit is 9.54 dB lower than its IP3 point. It should now be apparent why the IP3

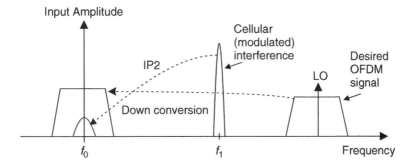

Figure 3.8 Example of modulated cellular blocker signal interacting with second-order nonlinearities of receiver generating components at baseband that can interfere with desired down-converted OFDM signal in direct-conversion receiver.

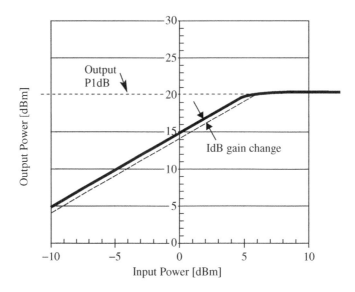

Figure 3.9 Example illustrating definition of 1-dB compression point (P1dB) of an amplifier.

point is only an extrapolated value. The signal compresses well before reaching this point.

3.4.2 Tools for Analyzing Modulated Signal Distortion

As illustrated above, IP2, IP3, and P1dB are useful tools for analyzing the linearity of the system in the presence of CW signals. These tools can also be utilized to get an idea of the behavior of the system in the presence of modulated signals. However, often more sophisticated tools are required that take into account the characteristics of the modulation such as the signal statistics and the PAR of the signal.

One such method of characterizing the linearity of a system is the analysis of the amplitude distortion (AD) and the phase distortion (PD) of the system. This method of analysis is often applied to linear power amplifiers which are required to amplify modulated signals with large PAR and simultaneous phase and amplitude information content such as WLAN OFDM. Although the derivation of the mathematical relation between AD, PD, and the statistics of the signal is beyond the scope of this text,[58] the understanding of some basic concepts such as AM–AM and AM–PM distortions as

[58]The interested reader, can, for example, see Kim et al. (2005).

well the cumulative distribution function (CDF) and the complementary CDF (CCDF) is quite useful and provides insight into the behavior of the system in the presence of such modulated signals.

AM–AM refers to the generation of undesired amplitude modulation due to the interaction of the input signal to the system with the nonlinearities of the system. AM–AM can result in gain expansion (where the measured gain is larger than the small signal gain) for a certain range of input signals and in gain compression (where the measured gain is smaller than the small signal gain) for other ranges of input signals. For a large enough input signal, all practical amplifiers will experience gain compression. An AM–AM plot can be very similar to a P1dB plot (Fig. 3.9). However, in order to analyze the system behavior over the large PAR experienced due to the modulated signal, the output signal level is often shown over a wide range of input signals and often well into the saturation power range of the system. Note that amplitude distortion is often referred to by system designers as *nonlinear distortion*.

AM–PM refers to the generation of undesired phase modulation due to the interaction of the input signal to the system with the nonlinearities of the system. AM–PM is particularly problematic in systems with high order constellations such as QAM-64 (which is utilized in 802.11a/g standards). AM–PM can occur, for example, when the capacitance of a tuned circuit varies as a function of the applied signal level. As a result, the output phase of the circuit would change nonlinearly as a function of the modulated signal. In WLAN systems operating at 54 Mbps QAM-64 modulation, the PD often limits the maximum transmit power before the AD. This is particularly true of tuned CMOS implementations due to their relatively large device nonlinear capacitances.

The CDF and CCDF are used to look at the probability of a certain peak (relative to the average) occurring in the signal. For example, a CDF plots the power level relative to the mean power level on the *x* axis, and the probability of its occurrence on the *y* axis. The plot of Figure 3.10 shows that the probability of a signal with a 15-dB peak relative to the average is close to zero even for an ideal signal. Nonlinear (amplitude) distortion typically causes clipping on the largest peaks of the signal. As a result, the distorted signal does not show as high of an occurrence for the largest peaks as an ideal OFDM signal would have (the CDF of the clipped signal falls inside that of an ideal signal). By comparing the CDF of an ideal signal with that of a measured signal, one can identify if amplitude distortion is causing degradation in EVM performance. Note that a CDF plot can identify both gain compression and gain expansion issues.

Figure 3.11 displays the AM–AM and AM–PM (at 5.24 GHz) for a commercially available class A power amplifier obtained with a VNA. Note that a

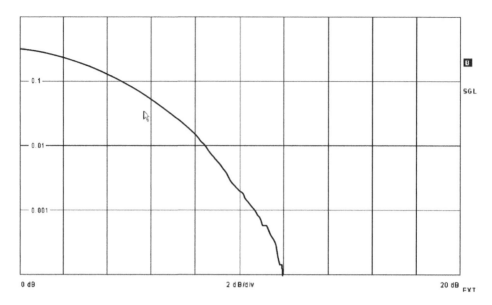

Figure 3.10 CCDF for almost ideal 802.11a OFDM signal with EVM of –45 dB and PAR of greater than 10 dB.

20-dB attenuator at the output of the power amplifier ensures that the measured AM–AM and AM–PM are not impacted by the nonlinearity of the VNA itself. It can be seen that a small signal gain of 22.7 dB is obtained (marker 1). The output 1-dB compression point of this amplifier is approximately 25.3 dBm and a phase distortion (deviation of the phase relative to the phase at small signal input) of less than 1° can be observed at this power level.

Figure 3.12 illustrates the CCDF of the amplifier of the same commercial amplifier under "lightly" and "heavily" compressed conditions. Under the light-compression condition, an output burst power of 17.3 dBm, a PAR of 7.5 dB, and an EVM of –25 dB are obtained. It can be observed that an 8-dB backoff from the P1dB of the amplifier is required to achieve this result. Under the heavy-compression condition, a burst power of +23 dBm, a PAR of 4.2 dB, and an EVM of –15 dB are obtained. Clearly, the EVM under the heavy-compression condition is not acceptable for the 802.11a/g standard for 54-Mbps transmission.

Figure 3.13 displays the spectral mask of the 802.11a standard applied to the output signal of the same commercial amplifier under the lightly and heavily compressed conditions corresponding to those of Figure 3.12. In the lightly compressed case, the amplifier easily passes the spectral mask requirements (this should not be surprising as the –25-dB EVM requirement of the standard is a more stringent requirement than that of its spectral

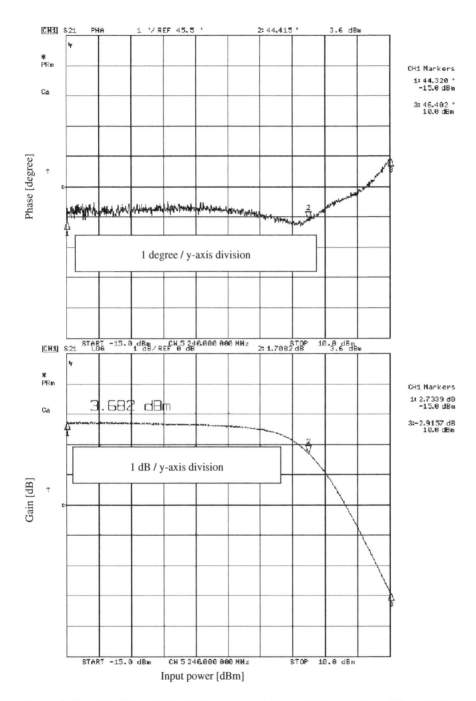

Figure 3.11 AM–PM and AM–AM for commercially available class A amplifier with 20-dB attenuator at output. This amplifier has an output P1dB of approximately 25.3 dBm (marker 2) and an AM–PM of less than 1° at this power level.

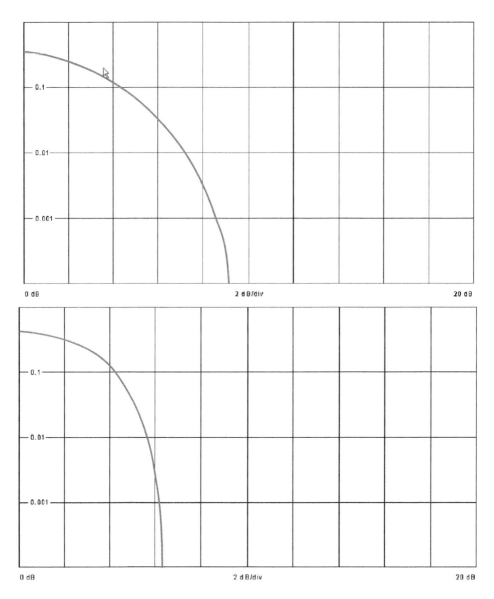

Figure 3.12 CCDF of amplifier of Figure 3.11 under "lightly" compressed and "heavily" compressed conditions.

mask). However, in the heavily compressed condition, the system fails the spectral mask requirements due to spectral regrowth associated with the excessive nonlinearities.

Note that the PA discussed here is a highly linear class A, laboratory-class amplifier. However, at the same time, power efficiency is not of much

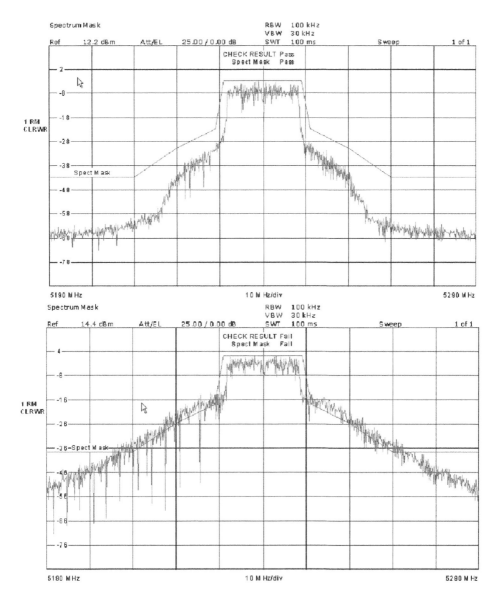

Figure 3.13 802.11a spectral mask of amplifier of Figure 3.11 under "lightly" compressed and "heavily" compressed conditions.

concern for this power amplifier which consumes 200 mA from a 15-V supply. For example, at 0 dBm out, this PA is only 0.03% efficient! At 17.3 dBm, the maximum 802.11a 54-Mbps compliant power level, it is still only 1.8% efficient. The challenge in the design of real WLAN PAs for portable applications is to design for these modulation types but to maintain a much higher level of efficiency. This, particularly in CMOS designs, is quite challenging and an area of active research.

3.5 RECEIVER IMAGE REJECTION

Another important radio impairment applicable to WLAN systems is the receiver image rejection. Image rejection is very closely related to quadrature balance (phase and amplitude balance), and therefore this topic will also be discussed in this section. It will also become apparent that many of the topics discussed here in the context of a receiver's quadrature balance are also applicable to a transmitter.

The system level issues related to image rejection are quite dependent on the chosen receiver architecture. Therefore we will look at the impact of poor image rejection on each of the discussed architectures separately.

3.5.1 Superheterodyne Receiver

In a superheterodyne receiver the RF mixer (Fig. 3.14a) translates the desired channel from the carrier frequency of ω_0 to an intermediate frequency of ω_{IF} (Fig 3.14b) by mixing the RF signal with a local oscillator signal at ω_{LO}. A standard (non-"image-reject" mixer) would however translate the frequency band at the "image frequency" $\omega_{IM} = \omega_{LO} + \omega_{IF}$ also to the same intermediate frequency. As a result the desired signal and the image signal would interfere with one another and would be indistinguishable.

The simplified block diagram of a superheterodyne architecture is shown again in Figure 3.14a. Typically one or both of the RF filters are used. RF filter 1 typically acts to select the desired band while suppressing the image channel, while RF filter 2 is typically used to reject the image channel further and also reduce the impact of image noise. Note that, given a high enough image frequency, a very simple filter can be used to attenuate the image noise by several decibels, resulting in up to 3 dB of improvement in system NF. Since RF filter 1 is placed before the LNA, it can contribute significantly to the NF of the system. However, in the absence of RF filter 1, the LNA can be driven to saturation by a large image or out-of-channel frequency. The decision on whether to use such a filter or not is therefore dependent on the system requirements and a trade-off between interference robustness and sensitivity levels.

In the absence of a lossless RF filter 2 and assuming a white output noise for the LNA, the overall system NF can be degraded by as much as 3 dB due to the noise of the LNA at the image frequency. This image noise will fall

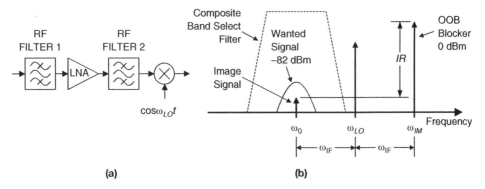

Figure 3.14 (**a**) Front end of direct-conversion receiver with filter stages identified. (**b**) Rejection of image component obtained using band-select filter.

on the desired band and add 3 dB to the noise level while not contributing to the desired signal level. In reality there is some noise associated with RF filter 2, the noise is not completely dominated by the LNA, and there is some gain roll-off at the image frequency of the LNA, so the actual improvement in NF due to the use of RF filter 2 is less than 3 dB.

This discussion is also relevant to the concept of SSB NF versus DSB NF of a mixer (or a whole receiver). A SSB NF is relevant when the desired signal occupies only one sideband around the carrier while the noise occupies both sidebands. A DSB NF is relevant when the signal occupies both sidebands around the carrier, and the noise occupies both sidebands around the carrier (or equivalently both the signal and the noise occupy one side of the carrier). Clearly, given infinite image rejection, SSB NF (dB) = DSB NF (dB) + 3 dB.

Figure 3.14 displays a hypothetical example of a large out-of-band blocker (OOB) with a 0-dBm signal power present at the image frequency of a desired signal with a −82-dBm power level (−82 dBm is the sensitivity level required by the 802.11a standard for a 6-Mbps signal). Assuming a required SNR of 5 dB for properly demodulating the 6-Mbps signal and that it is desirable to attenuate the interferer to the noise level, a minimum image rejection of 87 dB would be required. This 87 dB of rejection would have to be obtained from RF filter 1, RF filter 2, and any natural frequency selectivity of the LNA. From a rejection point of view it is preferable to have most of the rejection obtained through RF filter 1 so that the LNA is not exposed to large interference signals and would not run into compression. However, a higher rejection in the filter typically comes at the expense of larger in-band losses for the filter, which would directly impact the receiver NF and sensitivity. Therefore the rejection would have to be distributed between the two filters based on several factors, including the linearity, gain, and NF of the LNA.

It should be clear that, given a selection of front-end filters, the choice of IF has a significant impact on the image rejection of a superheterodyne receiver. A higher IF would relax the specifications on the RF filters (since the interference is farther away from the desired frequency). However, a choice of a higher IF would result in a higher Q requirement for the IF filters which is typically used for channel selection.

It is also important to note that, in reality, it is often required to have the image be several decibels below the noise floor. This is because the system would likely require an SNIR (signal to noise plus interference ratio) of –87 dB. This would require the image rejection to actually be a few decibels better than 87 dB in this example. In this example, a CW tone was used as an image to simplify the discussion. In reality the image is often a modulated signal. The magnitude of the required image rejection is also dependent on the type of modulation present at the image channel.

Note that sensitivity in the presence of an "interferer" as specified by the standard (e.g., adjacent channel interferer) does allow for the sensitivity level to be relaxed by 3 dB as compared to the case where there is no interference present. However, the standard does *not* allow for the relaxation of the sensitivity specification in the presence of an image.[59]

3.5.2 Low IF Architecture

In the low IF receiver architecture (Fig. 3.15a) the signal is applied to the LNA input with no RF filtering. The signal is then down converted through quadrature down converters. The image rejection is achieved through quadrature matching and complex bandpass channel-select filtering. As compared to a superheterodyne (high IF) architecture, the image rejection requirements are significantly reduced. This is because the image frequency is close in frequency to the desired channel. As a result the maximum power of the image frequency is typically controlled by the maximum ACI and AACI power allowed by the standard (in contrast to the superheterodyne architecture in which the maximum power of the image channel is set by a different standard or FCC limits and can be significantly larger than the desired channel). Note that since the maximum amplitude of the image is bounded by the standard, an RF filter is often not

---

[59]This is reasonable since the choice of the IF which determines the image frequency is not specified by the standard. It is a choice of the system designer. The ACI, on the hand, is specified by the standard and is not a function of the system design.

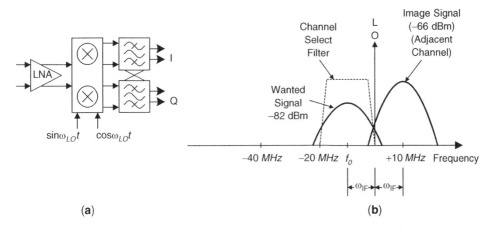

(a)   **(b)**

Figure 3.15 (a) Low IF receiver with baseband complex bandpass filters for image rejection. (b) Rejection of image component obtained using baseband complex bandpass filters and good matching in *I* and *Q* signal paths.

necessary for image rejection purposes (an RF filter would not be able to reject the image much in this architecture in any case, since the image is so close in frequency to the desired channel). An RF filter may still be required in order to select the desired band and avoid the saturation of the LNA by out-of-band interferers.

Once again, there is a trade-off between the choice of the low IF and the image rejection desired. A higher IF would typically result in a larger potential image channel power. At the same time, however, the required rejection (order and *Q*) of the baseband complex bandpass filters can be reduced. Note that in order to avoid signal aliasing, the choice of the low IF has to be at least as high as half the desired signal RF bandwidth. In the example shown in Figure 3.15b, the IF of 10 MHz is chosen, which is larger than 8.125 MHz (the RF bandwidth of the 802.11a signal is 16.25 MHz). It is also clear from this figure that the magnitude of the image rejection needed in the system is significantly less that that of the superheterodyne architecture. In this example, the same wanted 6-Mbps signal as the superheterodyne example is to be received. However, the image signal magnitude is governed and limited by the standard to –66 dBm. Using an analysis similar to that of the superheterodyne case, an image rejection of only 21 dB would be required. Further a typical LNA is not subject to saturation due to a –66 dBm signal.

The complex bandpass filtering and image rejection can be performed in the analog domain or in the digital domain. If the image rejection is performed in the analog domain, the dynamic range requirements of the ADCs would be reduced, as the stronger image signal is already rejected. If the complex bandpass filtering and channel selection are to be performed in the digital domain, the dynamic range requirements of the ADCs would be large enough to tolerate the large magnitude of the image signal while maintaining the integrity of the smaller desired signal. In this case, simple low pass filters can be used in the analog domain.

In quadrature-modulated, high order constellation, low IF transceivers, the quadrature accuracy requirements of the receiver may be set by the system EVM or PER requirements, which can be more stringent than that required for rejecting the image due to interference.

3.5.3 Direct-Conversion Receiver

As shown in Figure 3.16, in a direct-conversion receiver, the signal is received and passed on to an LNA with no image filtering. The signal is then down converted to baseband quadrature signals through a quadrature mixer. As seen in Figure 3.16b, in this case, no explicit image channel exists. Stated another way, the image of the desired channel is the desired channel itself. This is because the LO frequency is selected to be equal to the center frequency of the desired channel, and therefore the desired channel is placed on both sides of the LO. However, in order to maximize spectral efficiency, the data placed on the left side of the carrier are uncorrelated with that on the right side of the carrier. As a result a poor "image rejection" would result in spectral leakage and corruption of the received signal. In the case of

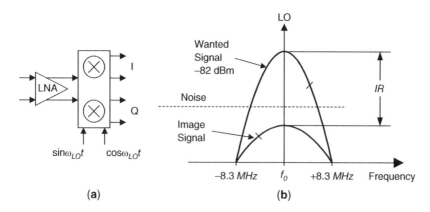

Figure 3.16 (a) Direct-conversion receiver with quadrature down converters. (b) Image of the signal is the signal itself. Good matching is still required on the *I* and *Q* paths to obtain a low EVM and high performance.

OFDM signals, in a receiver with no impairments, the subcarriers are mostly orthogonal to one another and therefore uncorrelated. As a result of a poor quadrature imbalance, the resultant image adds to the noise floor and distorts the OFDM constellation.

Although there are no explicit image rejection requirements for a direct-conversion receiver, it is still important to maintain good quadrature balance on a WLAN direct-conversion receiver. This is due to the fact that the OFDM high order constellations require a high accuracy quadrature signal to generate the EVM required to achieve the specified PER required by the standard. Intuitively it is clear that a higher order constellation (such as 64-QAM) would have much more stringent quadrature balance requirements than a low order constellation such as 4-QAM (also known as QPSK).

It is important to note that a poor IQ imbalance would not only adversely affect the sensitivity point of the receiver but also the PER floor.[60] As an example a receiver with poor quadrature imbalance could have a sensitivity of −70 dBm for a 54-Mbps data rate and a PER floor of 2%. If the IQ imbalance is removed from the same receiver, a sensitivity of −73 dBm and a PER floor of 0.1% can be achieved for a 54-Mbps data rate. This is because quadrature imbalance is a multiplicative impairment (in contrast to thermal noise, for example, which is an additive impairment).

3.6 QUADRATURE BALANCE AND RELATION TO IMAGE REJECTION

The discussions of quadrature imbalance so far were primarily concentrated on receiver effects, although many of the general concepts, especially in direct-conversion receivers, are applicable to the direct-conversion transmitters also.

The forthcoming discussions on quadrature balance are directly applicable to both the transmitter and the receiver.

Figure 3.17 displays the relationship between quadrature gain and phase error and image rejection. The x axis displays the phase error in degrees and the y axis the resultant image rejection for various amplitude errors. As an example, a phase error of 1° and an amplitude error of 1% would result in an image rejection of −40 dB. As will be seen shortly, in order to maintain a reasonable EVM for a 54-Mbps 802.11 signal, an image

---

[60]The PER floor is the (lowest) level of PER achieved when the incoming signal is approximately in the midrange of acceptable input powers such that the receiver PER is not impacted by low SNR or high levels of distortion.

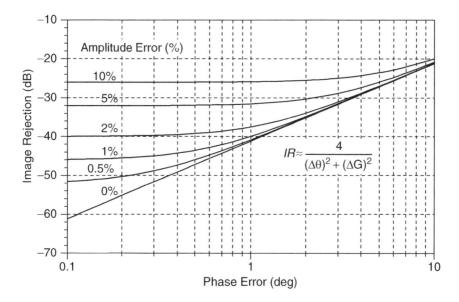

Figure 3.17 Relationship between phase error, amplitude error, and image rejection in quadrature modulator/demodulator. [Figure after Lee (1998)]

rejection of better than −32 dBc is desirable. Such an image rejection can result, for example, from a phase error of 3° and amplitude error of 2%.

For small errors, the magnitude of the image rejection can be approximated by the relation[61]

$$IR \approx \frac{4}{(\Delta\theta)^2 + (\Delta G)^2}$$

where $\Delta\theta$ is the phase mismatch in radians and ΔG is the gain mismatch in ratiometric terms. In terms of decibels

$$IR_{dB} \approx 10 \log\left(\frac{4}{(\Delta\theta)^2 + (\Delta G)^2}\right)$$

Note that in the chart of Figure 3.17 the x axis is on a logarithmic basis. It is not therefore surprising that for very low image levels a very small

[61]For a more detailed description of the exact relationship between phase and amplitude error as well as the derivation of the relation, the interested reader can refer to Lee (1998) or Behzad (1995).

change in phase error can result in a large change in the image rejection ratio.

3.7 QUADRATURE BALANCE AND RELATION TO EVM

As eluded to earlier, quadrature imbalance in a transmitter results in a finite rejection of the undesired sideband on a transmitter. This is similar to the effect that was described for a receiver, except that there are no interfering image signals to be concerned about. In the transmitter, the image rejection would be determined by the imbalance of the DACs, the baseband I and Q signal paths, and the LO quadrature accuracy. Quite often, especially in the case of a direct conversion transmitter, the dominant source of the quadrature imbalance is the I and Q LO signals. Maintaining quadrature accuracy at high frequencies is difficult due to very high sensitivity to parasitics and coupling paths. The parasitics involved can be electric (capacitive) or magnetic (inductive) in nature. In general, however, a quadrature inaccuracy due to the LO signals is (for all practical purposes) independent of the baseband frequency and can therefore be compensated by a fixed phase and amplitude predistortion term at baseband. A baseband-induced quadrature imbalance, on the other hand, is usually dependent on the baseband frequency and may require a more sophisticated correction. Such baseband quadrature mismatches usually require a constant group delay correction (linear phase with frequency) rather than a constant phase correction.

For an 802.11 OFDM signal, in general, the ith data subcarrier is uncorrelated with the $-i$th (data) subcarrier. The I/Q mismatches cause the interference between these mirror-image subcarriers resulting in the loss of orthogonality. Since in the ideal case the ith and $-i$th subcarriers are uncorrelated, the I/Q imbalance on an OFDM system exhibits itself as noiselike on the constellation diagram for the data subcarriers. This phenomenon is especially true for the higher order constellations.

On the other hand, the phase of the jth pilot subcarrier (transmitted in BPSK) is highly correlated with that of the $-j$th (pilot) subcarrier. As a result, I/Q amplitude imbalance results in distinct and non-noise-like constellation points on the constellation diagram across the real axis. At the same time, I/Q phase imbalance results in a distinct (non-noise-like) constellation point on the constellation diagram across the imaginary axis. The above statements can be verified by mathematically looking at the output generated from an amplitude mismatch,

$$C_{k,m}e^{j2\pi k\Delta ft} + \left(C_{k,m}\frac{\Delta G}{2}\right)(e^{j2\pi k\Delta ft} + e^{-j2\pi k\Delta ft})$$

or a phase mismatch,[62]

$$C_{k,m}e^{j2\pi k\Delta ft} + j\left(C_{k,m}\,\frac{\Delta\theta}{2}\right)(e^{j2\pi k\Delta ft} + e^{j2\pi k\Delta ft})$$

where $C_{k,m}$ represents a complex number corresponding to the location of the symbol in the constellation for the kth subcarrier and the mth symbol time. As before, $\Delta\theta$ is the phase mismatch in radians and ΔG is the gain mismatch in ratiometric terms.

By looking at the pilot tone constellation points, therefore, one can determine whether or not phase and amplitude imbalances exist in the modulation.

So, in summary, since the subcarriers are mostly uncorrelated, the image adds to the noise floor and distorts the OFDM constellation. The image also causes the orthogonality of the OFDM modulator signal to suffer, resulting in a degradation of the EVM.

The ideas of the previous section are elaborated using two examples. Figures 3.18a,b display the constellation diagram of an 802.11a signal which has been corrupted by *IQ* impairments. In the case of Figure 3.18a, the signal has been subject to a fixed phase error of 4° (and no amplitude error). As can be seen by looking closely at the vertical splitting of the pilot-based constellation points (shown in the boxes), one can deduce the type of impairment to which the signal has been subject. In Figure 3.18b, the signal has been corrupted by a 0.5-dB amplitude error (but no phase error). Again, by observing the constellation points associated with the pilot tones (horizontal splitting inside the two boxes), one can determine the type of impairment that has corrupted the desired signal. By observing these examples, it should be clear that making a conclusion about the type of impairment by looking at the constellation points associated with the nonpilot tone subcarriers would be difficult, especially when looking at a high order constellation diagram such as that of a QAM-64 modulation. The EVM per subcarrier of the signals of Figure 3.18 is plotted in Figure 3.19. Unlike some of the examples provided earlier, in this case, it is clear that by looking at the EVM–subcarrier plot, it is not possible to determine the cause of the signal impairment and the constellation diagram itself provides a better clue to the cause of the impairment. It is interesting to note that under this type of impairment (quadrature imbalance) the pilot subcarriers (subcarriers indexed –21, –7, 7, 21) show the highest level of EVM per subcarrier.

It should be clear that an OFDM constellation diagram with the constellation points for the pilot tones split at a diagonal angle (rather than strictly on

---

[62]For details see Cutler (2002).

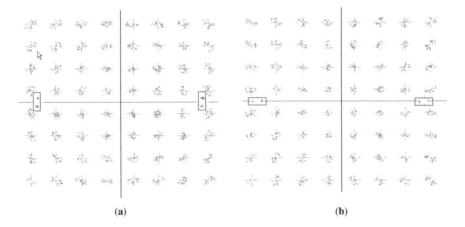

Figure 3.18   Constellation diagram of 802.11a signal with no impairments other than (**a**) a 4° phase error (EVM = –27 dB) and (**b**) a 0.5-dB amplitude error (EVM = –27 dB). Note that the characteristics of the constellation errors may be quite useful In debugging. Channel estimation based on *preamble only* is used on the VSA for these plots.

the imaginary or real axis) would have been corrupted by both phase and amplitude imbalance.

Figure 3.20 shows the same signal of Figure 3.18a, but with a change in the VSA settings. Instead of estimating the channel characteristics based on the preamble only, the VSA is now utilizing the information content in the payload also to estimate the impairments in the received signal. As a result it is now able to significantly improve the *IQ* balance for the pilot tones as well as to a lesser degree for the other subcarriers. This results in a significant improvement in the overall EVM of the received signal. In this case an EVM of –31 dB is achieved. Note that in this case the splitting observed in the constellation points associated with the pilots is no longer present. In order to utilize this debugging tool, therefore, the option for channel estimation based on the *preamble only* should be selected on the VSA.

It is also important to note that fixed phase errors (i.e., a phase error that is constant for all subcarriers ranging from low frequency to high frequency) are typically caused by an imbalance in the LO generation circuitry which is used to generate the *I* and *Q* outputs in the receiver and the single-sideband-modulated signal in the transmitter. On the other hand a subcarrier-dependent phase error (typically a constant delay) can be caused by mismatches at the baseband circuitry. Since a constant delay translates into a linear variation of phase as a function of frequency, the different index subcarriers would be subject to different amounts of phase error. Baseband-type mismatches are typically stronger closer to the filter edges (high pass

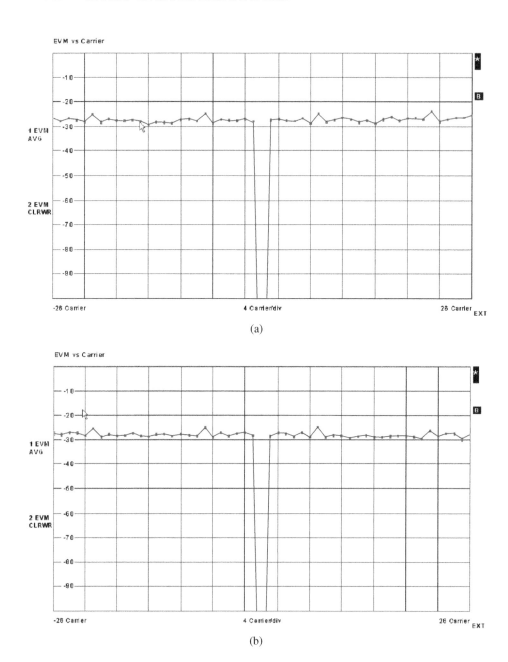

(a)

(b)

Figure 3.19 The EVM per subcarrier for signals of Figure 3.18: (**a**) 4° phase error; (**b**) 0.5-dB amplitude error. Channel estimation based on *preamble only* is used on the VSA for these plots.

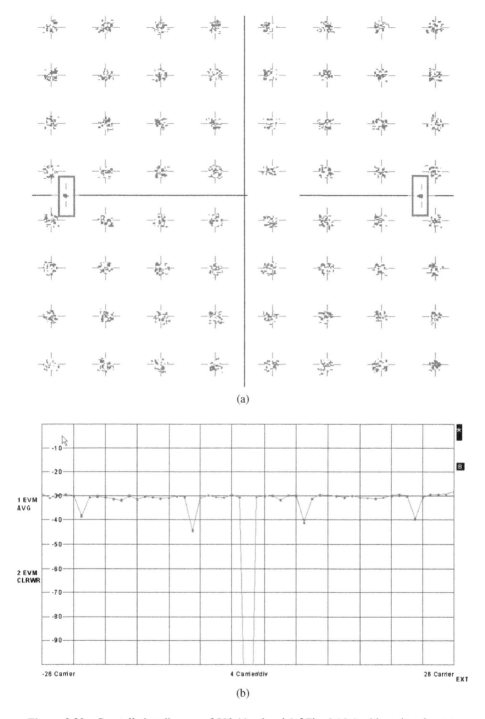

(a)

(b)

Figure 3.20 Constellation diagram of 802.11a signal (of Fig. 3.18**a**) with no impairments other than a 4° phase error (EVM = −31 dB). (**a**) Constellation diagram. (**b**) EVM vs. subcarrier. Channel estimation based on *preamble and payload* is used on the VSA for these plots.

corners as well as low pass corners) since small filter component mismatches can cause relatively large phase imbalance at the passband edges.

Quite often, in practical systems, the phase imbalance is of the constant type and is a result of imbalances at the LO generation circuitry. This is because, in contrast to the low frequency baseband circuitry, a small mismatch can result in a large phase imbalance at the RF. The higher the LO frequency where the I and Q signals are generated, the more difficult it would be to maintain perfect quadrature. As explained in the architecture discussions, this is one of the fundamental advantages of a superheterodyne design as compared to a low IF or zero IF design.

Subcarrier-dependent phase imbalance as well as subcarrier-independent phase imbalances can be corrected for using digital pre- or postdistortion. Subcarrier-independent phase distortion is somewhat easier to implement in the system. Amplitude imbalance can also be compensated in the digital domain, but one needs to be aware of the additional need for dynamic range on the ADCs and DACs and account for this in the system design.

At this point it would be natural to ask about the relationship between image rejection and EVM for a WLAN signal. Figure 3.21 displays this relationship based on a set of inputs generated by an arbitrary vector modulator and measured on a vector signal analyzer. All other impairments were disabled (limited by the quality of the vector signal modulator) and sufficient

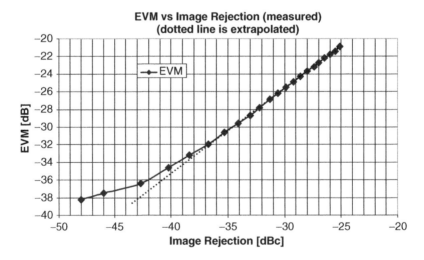

Figure 3.21 Measured EVM of "ideal" transmitter as function of its image rejection. The degradation of the EVM at low signal levels is due to the limitations of the VSA at low power levels. The dotted line represents the extrapolated value for an ideal VSA. Channel estimation based on *preamble only* is used on the VSA for these plots.

phase and/or amplitude imbalance was introduced in the modulated signal. The resultant EVM was measured and plotted on the y axis. On a log–log plot as shown (both x and y axes are in decibels), EVM (decibels) degrades almost "linearly" with image rejection (decibels). The departure of this linear relationship (shown by the dotted line) at lower image rejection levels is due to the limitations of the vector signal modulator and the vector signal analyzer at these low EVM levels.

As an example, using this figure, it is clear than in order to satisfy the EVM requirements of a 54-Mbps 802.11a signal, the image rejection of the system is required to be better than approximately –29 dBc. Using Figure 3.17, it can be seen that in order to achieve such an image rejection, an amplitude error of 5% and phase error of 2% can be tolerated. Note that in this example it is assumed that no other impairments exist in the system. In reality this is never the case, and therefore a system budget is required which determines how much imbalance can be handled in the system in the presence of other impairments. In the case of a transmitter, often the EVM is (and should be) limited by the nonlinearity of the power amplifier. It would therefore be desirable to degrade the EVM due to all other impairments quite minimally. Therefore, a worst-case image rejection of better than –35 dBc is quite desirable.

For the newer emerging standards such as 802.11n, even better EVM numbers will likely be required by the system. As such the *IQ* imbalances need to be proportionally smaller. By looking at Figure 3.21, it is clear that, for example, an EVM of –30 dB would require an image rejection of approximately –34 dBc (assuming that all the error budget is allocated to the quadrature imbalances). In reality an image rejection of better than –40 dBc would be highly desirable.

Achieving an image rejection of about –40 dBc over process, temperature, supply voltage, and channels, especially for the 802.11a standard (which spans 4.9 to 5.8 GHz), is quite challenging. As a matter of fact, achieving such performance with no calibrations in the system while achieving high production yields may be impossible. Various calibration schemes have been devised in order to enable such high degrees of image rejection. Some of these techniques will be discussed in some detail in Chapter 5.

The results of Figure 3.21 were obtained with VSA tracking enabled during the preambles only. If tracking is enabled during the preambles *and* payload, better EVM results will be obtained for the same image rejection (and quadrature imbalance). This improvement is typically around 2 dB but in some cases can be larger. In other words, the same radio will produce better system results (in this case EVM and PER) when it utilizes a better digital PHY. This points out the fact that it may be difficult to compare two radios based on system results without considering the baseband unit which is used in conjunction with the radio to obtain the system results.

3.8   OTHER TRANSMITTER (MODULATOR) IMPAIRMENTS

In addition to IQ impairments which can limit the signal quality of a transmitter, the designer has to be aware of a variety of other impairments that can impact a modulator.

Carrier leakage (also known as LO feedthrough, or LOFT) is one such impairment. LOFT is typically caused by two very different sources.

The first source of LOFT is DC offsets at the baseband path. It is easy to see that with baseband offsets the output-modulated signal would be given by

$$\text{Out}(t) = [A(t) + V_{OS}] \cos[\omega_{LO}t + \phi(t)]$$

The output-modulated signal will therefore possess an undesired carrier feedthrough at the LO frequency. It should be clear that from the point of view of LOFT a larger signal throughout the baseband section is highly desirable. This desire, of course, places a stringent requirement on the linearity of the baseband chain. As an example, in order to achieve better than 30 dB of carrier suppression due to baseband offsets, the baseband offsets (V_{OS}) must be 30 dB smaller than the RMS value of the modulated signal $A(t)$. In the simple example of a single-sideband sine wave output (with sine and cosine baseband inputs) with 1 V_{pp} amplitude on each quadrature rail, the input offsets on each rail will need to be no more than

$$\frac{1/2\sqrt{2}}{20 \log(30/10)} = 37 \text{ mV}$$

Note that in the case of a modulated signal with a large PAR such as that of the 802.11a/g the situation is worse. This is because in order to maintain good linearity in the DAC and the analog and RF chains, a significant amount of backoff from the clipping level of the circuitry is required. This results in a decrease in the RMS level of the signal, while the offset levels do not scale (decrease). For example, if the DAC and the analog/RF chain is designed to operate with a 15-dB backoff from the clip level of the DAC, and this level is assumed to be 1 V_{pp}, to obtain better than –30 dBc LOFT, the DC offsets at the input to the modulator (mixer) have to be smaller than

$$\frac{1/10^{15/20}}{20 \log(30/10)} = 18.6 \text{ mV}$$

It is important to note that in the case of a quadrature modulator each quadrature rail needs to have no more than this amount in DC offsets. In the case of a perfect quadrature modulator (with zero or very little image), an equal amount of α DC offset on each of the I or Q rails would result in 3 dB worse LOFT in decibels relative to the carrier than if only one of the rails had the same amount of (α) DC offset.

A second fundamental mechanism for generating LOFT at the output of the modulator is through parasitic coupling of the LO signal (needed for up conversion of the baseband signal) to the RF output port of the mixer or possibly even stages following the up-conversion mixer. To reduce the LOFT due to this mechanism, the parasitics should be kept to a minimum, and for differential circuits the balance should be maintained to the extent possible. Since parasitic couplings are more prevalent at higher frequencies, it is natural to expect more LOFT due to this mechanism at higher frequencies.

For direct-conversion transmitters, the single LO signal would directly couple to RF and show up at the center of the band of the transmitted output. For superheterodyne transmitters, the LOFT of the first LO would couple to the IF output (of the first mixer) and then unconverted to the center of the RF band by the second mixer. The LOFT due to the second mixer would typically fall out of band and is easily filtered by tuned circuits or external filters. The LOFT is typically a bigger problem for direct-conversion transmitters since, as mentioned earlier, parasitic couplings would be larger at the RF as compared to that of the IF of the superheterodyne transmitter.

If the desired signal contains information content at DC, LOFT caused by DC offsets or direct RF coupling would corrupt the modulated signal and can be detrimental to the system performance. If the desired signal has no information content at DC, LOFT is typically not detrimental in the system performance but for a variety of other reasons should be kept to a minimum. One reason for the desire to keep LOFT to a small value on the modulator is that in the absence of frequency offsets in the system (i.e., perfectly tuned crystals) a large LOFT can cause large DC offsets on a direct-conversion receiver, thereby limiting the system performance.[63] In the case of the 802.11a and 802.11g standard, the OFDM-modulated signal nulls out the zero-order subcarrier and therefore does not possess DC information content. However, due to reasons explained earlier, the 802.11 standard imposes guidelines on the maximum allowed LOFT power relative to the desired signal power (−15 dBc).

---

[63]This is because the large incoming signal is further amplified by the LNA and applied to the mixer RF port. A large LO signal also exists on the mixer LO port. These signals can mix and create a large DC offset.

Often it is necessary to provide gain control in the transmitter. Gain control would be needed, for example, to compensate for transmit power variations due to changes in temperature, process, or supply voltage.[64] Further, many standards, including 802.11, may require transmit power control in order to increase the overall capacity of the cell and reduce the near–far problem.[65] Another obvious reason for the use of power control is to reduce power consumption of the transmitter when the distance between the transmitter and receiver is short.

As a result of these requirements, often tens of decibels of power control may be desired in the system. Gain control can be achieved at baseband or at RF or in the combination of the two. There are several system level trade-offs between the two methods of gain control. Typically, gain control at baseband allows for very accurate control of the gain steps as well as fairly insensitive results to process, voltage, and temperature variations. One way to achieve such insensitivity is through the use of a switched resistor divider (PGA) circuit prior to the up-conversion mixer. In such a scheme, the gain is set by the ratio of two resistors of the same type. An alternative baseband gain control scheme that is often used (but does not result in as accurate of gain steps) is through programmable degeneration in the transconductor of a mixer.

However, often, gain control at baseband results in poor LOFT results at high attenuation settings. Typical implementations of baseband gain control described above would attenuate the signal well in advance of the mixer switches. The mixer switches therefore observe a small signal but a fairly large DC offset, resulting in a significant reduction in the LOFT performance at high attenuation settings. It is therefore desirable to implement the baseband gain control as close as possible to the up-converter switching devices so that as gain is reduced the offsets are proportionally reduced also, and LOFT in decibels relative to the carrier remains approximately constant. An interesting method of a baseband gain control scheme that allows for good LOFT results even under high attenuation settings is discussed in Chapter 6.

If gain control is implemented in the RF section of a transmitter after the mixer block, any LOFT due to DC offsets would scale with the gain of the RF chain. As such, maintaining a constant dBc number should be relatively

---

[64]In a CMOS process, for example, gain of tuned RF blocks can change significantly. This is primarily due to two reasons: First, the transconductance of the CMOS device degrades significantly with increasing temperature (due to a reduction in the mobility of the device) and, second, the Q of on-chip inductors decreases significantly as a function of temperature due to an increase in metal resistivity at higher temperatures.

[65]In order to minimize the interference of one user with another, each user should use only the minimum amount of power needed to establish his or her link with the desired SNR.

simple. LOFT due to direct coupling of the LO signal to the mixer output would also scale with gain control. This scaling of LOFT with gain control is one of the major advantages of gain control at RF. It is important to note, however, that if there is a significant coupling path to a point beyond the RF output of the mixer, then the LOFT will no longer necessarily scale with RF gain. Traditionally this has not been a problem since the magnitude of such coupling has been small. However, as radios are reduced in size to reduce cost, undesirable electric and magnetic coupling is increased and can result in such cases.

In general, achieving the desired LOFT requirements of many standards over process/voltage/temperature and over gain control without some form of autocalibration is quite difficult. Many autocalibration schemes have been used to alleviate this problem. Some of these methods will be discussed in Chapter 5.

Another class of impairments associated with the modulator are those related to HDs of the baseband blocks. In a WLAN 802.11a or g direct-conversion transmitter, for example, the baseband block is typically designed with a (low pass filter) bandwidth of 8 to 10 MHz (so that an RF channel bandwidth of 16.25 MHz can be accommodated). So it is easy to see that the various harmonics of a subcarrier would cause interference with the subcarriers at the harmonic frequencies. Since the subcarriers are designed to be orthogonal with one another, this HD-related interference would cause inter-subcarrier interference and a resultant degradation of the EVM. It is noteworthy that the lower index subcarriers have a greater chance of causing EVM degradation due to harmonic distortion since their higher order harmonics also fall in band. As an example, theoretically the subcarrier at 937.5 kHz can have its harmonics up to HD8 interfere with the desired in-band signal, whereas the subcarrier at 6.25 MHz will cause no HD-related degradation to EVM (assuming a high enough ADC sampling rate to avoid aliasing).

It is important to note that the HD related to the RF blocks (blocks after the mixer) typically does not cause EVM degradation since these harmonics fall far out of band. These harmonics can, of course, cause FCC violations and need to be accounted for in the design.

A closely related impairment to HD is IM. As discussed earlier in this chapter, for a simple nonlinearity, a mathematical relationship can easily be derived relating the nth-order HD to the nth-order IM. Odd-order IM terms often fall very close by (in the spectrum) to the fundamental tones that are causing them. Therefore these IM terms are problematic for low order subcarriers and high order subcarriers at baseband as well as all the RF blocks of the transmitter.

CW tones at baseband can be used to identify many of a modulator's impairments. Figure 3.22 displays the output of a modulator as obtained on a

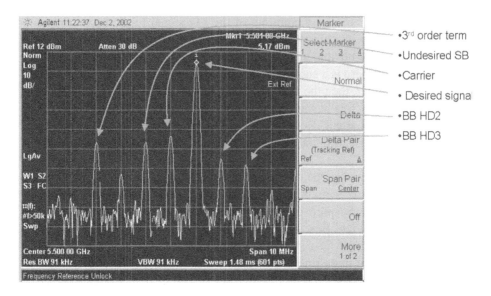

Figure 3.22 Baseband 1-MHz sine and cosine signals are applied to the quadrature input ports of a modulator running with an LO of 5.5 GHz. Several transmitter (modulator) impairments are identified in this plot.

spectrum analyzer when 1-MHz sine and cosine waves are applied at the quadrature and in-phase baseband input ports of the modulator. With the LO frequency set to 5.5 GHz, the following tones associated with the various impairments can be observed:

Desired tone	5.501 GHz
LOFT	5.500 GHz
Image tone	5.499 GHz
Baseband HD2	5.502 GHz
Baseband HD3	5.503 GHz

Additionally tones at 5.498 and 5.497 GHz are observed. These are associated with the images of the baseband HD2 and HD3 terms, respectively. It is interesting to note that the magnitude of the image tone of the HDn component can be larger or smaller than that of the HDn component itself. This is because the HDn terms on the I and Q rails do not have the same phase relationship as the fundamental tones. Therefore, when they are combined in a quadrature modulator, the magnitude of the image can increase or decrease. Stated differently, having a "perfect" quadrature only guarantees than no image component for the fundamental would exist. It does not guarantee, however, that no image components for the HD terms would exist. The best way to reduce the magnitude of the image components for the HD terms is to reduce the magnitude of the HD terms themselves by making the modulator more linear.

3.9 PEAK-TO-AVERAGE RATIO AND RELATION TO LINEARITY AND EFFICIENCY

Now that we have studied some of the impacts of nonlinearities in the transmitter (modulator), it would be important to discuss the interrelationship between linearity of the transmitter, PAR of the modulated signal, and the obtained system power efficiency.

In general, a signal with nonconstant (varying) amplitude complicates the transmitter design. One degree of the complexity of the transmitter design would depend on the extent by which the amplitude is varying. The extent by which the amplitude varies can be quantified by a concept known as peak-to-average ratio, or PAR. The system performance will depend on the PAR, the probability distribution function of the occurrence of the peaks, as well as the "linearity" of the transmitter. The linearity of the transmitter will not only be characterized by AM–AM (amplitude distortion) but also by AM–PM (phase) distortion.[66]

Of course, on the positive side, modulated signals with nonconstant envelope (which carry information in the phase as well as the amplitude of the transmitted signal) allow for much higher spectral efficiency.

Note that signals with information content in the amplitude domain are not the only ones requiring linear amplification (although they are often the most demanding ones). Signals such as QPSK which carry information in the phase domain but need to be limited in bandwidth due to practical reasons also require somewhat linear amplification to avoid spectral regrowth. This is because band limiting a signal with abrupt phase transitions such as QPSK introduces variation in the envelope, thereby requiring linear amplification.[67]

As discussed earlier, an OFDM-modulated signal is a nonconstant-envelope signal. As a matter of fact, the OFDM signal used in 802.11a/g has a theoretical maximum PAR of approximately 17 dB. In other words, the peak voltage excursion of an 802.11a/g signal can be seven times larger

---

[66]At very high power levels approaching the saturation power of the power amplifier, additional effects such as PM–AM, PM–PM, memory effects, and nonsymmetry of IMD products also have to be taken into account, at times, even for "constant-envelope" modulations.

[67]For a brief discussion on this topic see Razavi (1998). For more details refer to Sevic et al. (1996).

than that of its average signal. To illustrate this fact, Figure 3.23 shows an 802.11a/g-modulated signal payload in the time domain.

To prevent distortion and to be able to reproduce the amplified output signal faithfully, the transmitter would need to avoid any clipping even during the peak excursions of the signal (which do not occur very frequently). In order to achieve this, the PA would be designed to have minimal compression at the peak power. However, most of the time, the PA would be transmitting a signal that is seven times (17 dB) smaller than the peak power. If we assume that we would allow 1 dB of compression at the peak signal level, this would result in the transmitter operating at approximately 17 dB lower power than its 1-dB compression point. An inductively loaded class A transmitter and power amplifier can achieve a maximum power efficiency of 50% (achieved when transmitting the maximum output swing). Therefore the best case achieved efficiency would be

$$\eta < \frac{P_{\text{out}}}{P_{\text{dis}}} = \frac{50}{7} \approx 7\%$$

The inverse relationship between the amount of backoff from the 1-dB compression point (typically simply referred to as "backoff") required in an amplifier and the maximum achievable efficiency should be clear. Since the power consumption of a transmitter is often dominated by the PA at the output of the chain, it is desirable to be able to achieve the highest possible efficiency for this block.

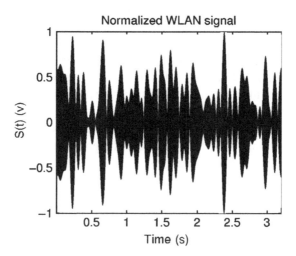

Figure 3.23 Time-domain waveform of 802.11a signal normalized to magnitude of 1. Notice the large PAR of the signal.

In reality, some clipping can be tolerated while still satisfying the requirements of the 802.11a/g EVM even at the highest data rate of 54 Mbps. In order to achieve the highest system efficiency, it is common to operate the PA in class AB mode and back off by about 7 dB from the 1-dB compression point.[68] The stages prior to the PA are typically operated at a larger backoff to make sure that the nonlinearity is not dominated by these stages. Since these stages burn a relatively small power, this trade-off allows for achieving the highest possible overall efficiency.

As mentioned earlier, given the high order QAM constellation used in 802.11, both AM-to-AM and AM-to-PM distortions have to be considered in evaluating the system performance. Conventionally, only the 1-dB compression point which only covers issues with AM–AM has been used to determine the linearity of power amplifiers. This may be due to the fact that older systems rarely use high order QAM constellation in which phase linearity and error are quite important. However, for high order QAM modulations, AM–PM will start degrading the system performance (as measured by EVM) well before the AM–AM (P1dB) effects are visible. Further, two PAs may have the same P1dB point but very different AM–PM characteristics (e.g., one may have a hard clipping and another a soft clipping). These are the primary reasons as to why a one-to-one correspondence between P1dB (or even IM3) and EVM performance does not exist. In order to completely characterize the performance of a power amplifier for 802.11, either EVM must be specified or plots for both AM–AM and AM–PM must be provided.

Power amplifier linearization and efficiency enhancement techniques are very important areas of research. Various techniques have been proposed that use analog or digital techniques to linearize the phase and/or amplitude of the power amplifiers. These techniques can be analog and/or digital, can be open loop or closed loop, and can use feedforward or feedback techniques. In order to achieve best linearity and obtain high efficiency, the P1dB of the amplifier should be increased using these techniques and the amount of backoff required to achieve a particular EVM should be reduced.

It is important to note that lower order 802.11 modulations such as the BPSK used for 6-Mbps transmission are much more tolerant of signal clipping and nonlinearities in the system. As a matter of fact, from an EVM and PER point of view, a receiver can still receive and achieve a 10% PER even if the transmitted signal was operating at or even passed the P1dB of the transmitter. On a transmitter for 802.11a data rates below 24 Mbps, the

---

[68]It is *very* important to note that the exact amount of backoff is very much related to the amplitude and phase nonlinearity characteristics of the particular PA in use. The 7-dB backoff mentioned above is only a "typical" number often utilized for common SiGe and GaAs PAs in the market today.

maximum transmit power is typically set by the spectrum mask, whereas for the higher order data rates, it is set by the EVM. More will be discussed on this topic later.

3.10 LOCAL OSCILLATOR PULLING IN PLL

We will now discuss impairments that impact the performance of the PLL.

The first impairment we will discuss is the "PLL pulling" (or VCO pulling) phenomenon. In the most general definition, PLL pulling refers to a phenomenon by which the LO generated by the PLL and supplied to the receiver or the transmitter momentarily changes frequency as a result of a transient event. Such pulling usually occurs by the transient event coupling to some part of the VCO (control line, power supply, etc.). Modern integrated VCOs usually have a very high gain (in MHz/V) due to a limited power supply headroom and the need for wideband tuning. The problem is therefore exacerbated in the newer generation of ICs unless proper techniques are utilized. If such a pulling is reasonably small, the receiver's baseband digital PLL may be able to correct for it, but if the frequency shift is large enough, the system can experience a packet loss.

One technique to reduce the effect of VCO pulling would be a mechanism to reduce the VCO gain (K_{vco}) by using banks of switch capacitors (to be discussed later). For a typical 802.11a worldwide system, without the use of such a technique, K_{vco} can be as large 1 GHz/V! With such a large gain, a transient noise event of 100 µV can generate a momentary transient frequency shift of 100 kHz. In a typical 802.11 system such a large frequency shift would be quite difficult to track by the receiver baseband PLL. By using switch-capacitor banks, K_{vco} for this system can be reduced to 30 MHz/V such that the same 100 µV noise event would result in a small 3-kHz transient frequency shift.

One of the most common reasons for VCO pulling is a phenomenon known as injection locking. In general, small perturbations to a VCO can be seen as phase noise or spurs. But as the spurs get closer to the VCO's operating frequency, and as they get larger in amplitude, injection locking occurs and the VCO tries to track the frequency of the coupled noise signal (Fig. 3.24a). As shown in Figure 3.24b, if the VCO frequency is close to the transmitted frequency (which can be the case in a conventional direct-conversion transmitter) and especially if a high power amplifier is integrated on the same die, the PA output can couple to the VCO and cause injection locking. Injection locking can be detrimental to the operation of a system and will result in extremely poor EVM. In general since the VCO has the highest gain at the frequency of operation, the closer the

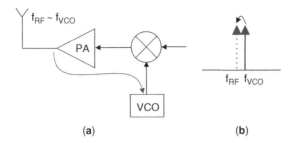

Figure 3.24 Example of VCO pulling caused by coupling of PA output to VCO. In this example, it is assumed that the VCO is operating at the center frequency of the PA output (as would be the case in a direct-conversion transmitter that uses no special provisions to avoid this situation).

interfering signal is to the operating frequency, the higher the chance of injection locking.[69]

Another mechanism that can cause VCO pulling is the reception of a very large interferer by the receiver. Again, in a direct-conversion or low IF receiver, the large interfering signal can injection lock to the VCO through parasitic coupling paths. Typically this is a smaller problem than the one described in the previous paragraph in a transmitter, because of the larger amplitude associated with the power amplifier output.

In both cases described above, one solution would be to operate the VCO at a different frequency than the transmitter (TX) or receiver (RX). The VCO can, for example, operate at half the TX output frequency or twice the output frequency. The LO is then generated by the proper integer division or multiplication function. Even a better solution would be if the VCO operates at a frequency that is not harmonically related to the transmit or receive frequencies. For example, the VCO can operate at two-thirds the RF. The advantage of this scheme is that no harmonics of the VCO would fall on the fundamental of the RF frequency and vice versa. Of course, there is a common frequency at which the nth harmonic of LO and the mth harmonic of the RF signal would both fall on, but this is rarely a problem.

So, in summary, listed below are some of the primary factors determining the magnitude of the TX or RX VCO pulling problems discussed above:

1. The magnitude of the interfering signal

---

[69]R. Adler first discussed injection locking in his paper, "A Study of Locking Phenomena in Oscillators," in 1946. Since then much work has been done to study this phenomenon. In many places injection locking is to be avoided and the proper precautions are to be taken. In other cases, injection locking is utilized for various purposes in electronic as well as optical systems.

2. The isolation from the interfering source to the VCO
3. The frequency offset of the interfering signal relative to the VCO frequency
4. The point of coupling on the VCO circuit
5. The Q of the tank

The coupling from the interfering source to the VCO can take many paths such as the magnetic or electric coupling on the package, magnetic or electric coupling due to on-chip components, and coupling through the chip substrate.

Another mechanism that can cause VCO pulling is the change of an impedance as seen by a VCO (commonly known as load pulling). For example, in a half-duplex system, as the system is switched from RX to TX or vice versa, different LO buffers may be turned on or off. This transient event often causes a transient change of load impedance. If this change of impedance is buffered from the VCO by a high reverse isolation (through the use of multiple buffers, cascode devices, etc.), then there will be little or no VCO pulling. Otherwise VCO pulling could occur.

In all of the above examples of VCO pulling, a wide bandwidth PLL can help recover from the pulling effect quickly. So in general a wide loop bandwidth would be beneficial. However, the loop stability is quite important too, since an underdamped loop (even if wideband) can cause frequency ripples in the PLL response which may degrade the system performance. One can observe the frequency-settling behavior of a PLL using a signal analyzer with an FM demodulator option. An example of the PLL settling behavior is shown in Figure 3.25. In this example the PLL settles to within 50 kHz of its final value in about 32 μs.

3.11   PHASE NOISE IN PLL

One of the most important specifications for a PLL is its phase noise performance.[70] Phase noise can be qualitatively described as the random fluctuations in the instantaneous phase as a result of noise altering the zero-crossing point in time.

Ideally an LO signal generated by the PLL would only have a frequency

---

[70]Phase noise is the most common specification for LO synthesis PLLs. Baseband PLLs, on the other hand, which are used for clock recovery, are typically specified in terms of their RMS and peak-to-peak jitter performance. Jitter is a time-domain specification (specified in, e.g., picoseconds), whereas phase noise is a frequency-domain specification (specified in decibels relative to the carrier per hertz (dBc/Hz) at a certain frequency offset from the carrier).

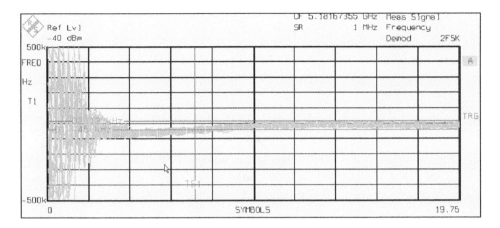

Figure 3.25 A PLL settling behavior obtained by using an FM demodulator on a VSA. The PLL settles to within 50 kHz of its final value in about 8 symbols or 32 μs.

content at the desired synthesized frequency and should look like a single "stick" when observed with a spectrum analyzer. In reality, the LO signal would have "side skirts." These side skirts result in the presence of energy at adjacent frequencies to the LO signal (Figure 3.26b).

The closed-loop phase noise behavior of a PLL is the result of the many components of the PLL contributing the overall phase noise with their unique transfer functions. For example, in the closed loop of the PLL, the phase noise of the VCO appears with a high pass transfer function.[71] In other words, although the VCO itself has a significant amount of noise at low frequency offsets from the carrier, this phase noise is rejected by the magnitude of the PLL gain at the lower frequencies. On the other hand, components such as the crystal oscillator, the reference dividers, the phase-frequency detector, and the charge pump exhibit a low pass noise transfer function as seen at the output of the PLL.

Based on the above description, it should be clear that a trade-off exists in the PLL phase noise performance relative to the loop bandwidth of the PLL. For example, if the loop bandwidth is set to a wide value, the VCO phase noise would be rejected to a high degree, but there would be a higher contribution to the integrated phase noise of the PLL due to the in-band noise contributors. Setting the loop bandwidth to a narrow value would cause the VCO phase noise to dominate the integrated phase noise performance. Note that in a fractional-*N* PLL, the quantization noise of the

---

[71]This can easily be seen by analyzing the loop equations to obtain the proper transfer functions as seen at the output of the VCO (which is the output of the PLL).

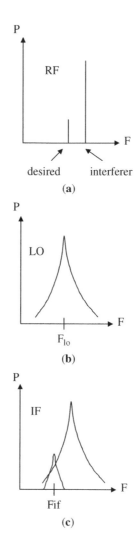

Figure 3.26 Impact of phase noise on reciprocal mixing in a receiver: (**a**) spectrum of desired signal and interferer modeled as CW signals for simplicity; (**b**) spectrum of nonideal LO which displays effect of phase noise; (**c**) output of receiver mixer after mixing spectrums shown in (**a**) and (**b**).

sigma–delta modulator would be quite large and should also be taken into account in setting the PLL BW. The narrower the PLL bandwidth, the more of the sigma–delta noise would be rejected by the loop action. This is traded off against the impact of the loop bandwidth on the VCO noise.

It is important to note that many other factors enter the equation in setting the PLL bandwidth. These include the PLL settling time, reference spurs, and loop stability. All these factors need to be accounted for in designing a PLL.

PLL and VCO phase noise can adversely affect the performance of a transceiver in several ways. Different standards would be more or less sensitive to degradations due to certain aspects of phase noise, and the PLLs would therefore be designed with these sensitivities in mind.

The general impact of phase noise in the system can be typically analyzed by looking at the "close-in" phase noise and "far-out" phase noise. Close-in phase noise would refer to phase noise within the bandwidth of the desired *channel* (not to be confused with the loop bandwidth of the PLL, which is typically set to a fraction of the channel bandwidth). Far-out phase noise can be described as the phase noise outside the desired channel bandwidth. In the case of 802.11a, with this definition, the close-in phase noise would refer to a single-sided phase noise of up to 8.125 MHz from the carrier frequency.

In the case of a receiver phase noise (typically out-of-band phase noise dominated by the VCO) can result in a phenomenon called *reciprocal mixing.* Consider the scenario in which a receiver receives two signals at its antenna: one desired small signal and an undesired large interfering signal reasonably close in frequency to the desired signal (Fig. 3.26a). The LO has a significant amount of phase noise on it (Fig. 3.26b). When the two signals are mixed at the mixer, the synthesized LO's noise is superimposed on both of the down-converted signals. At the output of the mixer, the small desired signal is now corrupted by the LO's noise which has been down converted by the large interfering signal (Fig. 3.26c). This mechanism is referred to as reciprocal mixing, because the RF port of the mixer has acted like an LO port (with the large interferer being the LO signal) and the LO port has become the RF port (with the LO noise becoming the "RF signal").

In a transmitter, LO phase noise creates a noisy output signal that can interfere with other users (such as the user in the adjacent channel) in the form of spurious emissions. The spurious emissions are typically restricted by the standard or by the FCC. In full-duplex systems such as WCDMA, the transmitter phase noise is typically highly restricted in order to reduce its impact on the receive band incoming signals. These restrictions are necessary due to potential for a very large dynamic range difference between the transmitted signal (e.g., +26 dBm) and the received signal (e.g., –98 dBm).

Reciprocal mixing and spurious emissions are examples of issues related to far-out phase noise. These can be very important, for example, in embedded WLAN applications where the WLAN radio is integrated along with the cellular radios in a cell phone.

In many applications of 802.11, however, the most important issues (and the most challenging specifications) related to phase noise are related to the close-in phase noise.

3.12 FAR-OUT PHASE NOISE

It is instructive to investigate the impact of far-out phase noise on system behavior. As discussed earlier, far-out phase noise can cause adjacent channel interference and reciprocal mixing. In order to avoid desensitization due to reciprocal mixing, the system phase noise has to be designed with a low enough out-of-band phase noise. We now consider an example.

The standard sets the sensitivity level requirement for the 6-Mbps data rate of 802.11a at –82 dBm. Assuming an SNR of 5 dB is required to properly decode this signal, we require that the products of the reciprocal mixing are below this level so that they do not impact the signal to noise plus interference level (SNIR) at the receiver slicer (Fig. 3.27). Let us further assume an adjacent channel CW interferer with –66 dBm power at a 12-MHz frequency offset from the center of the desired signal. Given the 16.25-MHz bandwidth of the desired signal, the required interference level due to reciprocal mixing would need to be less that –87 – 10 log(BW = 16.25 MHz) = –159.1 dBm/Hz. Given a –66-dBm interferer, the phase noise at an offset of 12 MHz would have to be smaller than –159.1 – (–66) = –93.1 dBc/Hz. Similar calculations show required PN of –109.1 dBc/Hz at a 32-MHz offset and –129 dBc/Hz at a 50-MHz offset, given the assumed interferers

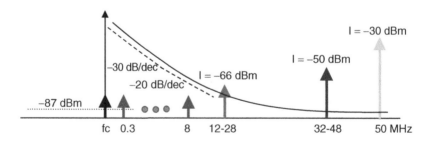

Figure 3.27 Impact of phase noise on reciprocal mixing in 802.11a receiver. The in-band subcarriers from 300 kHz to 8 MHz are shown. Assumed CW interferers are also shown at various offset frequencies. The profile of an assumed phase noise is also shown. Such phase noise often has –30 dB/decade at small frequency offsets and –20 dB/decade further away.

shown in Figure 3.27. Note that the CW ACI interference at a 50-MHz off-set is a Hyperlan II requirement and not an 802.11 requirement.

In reality, the adjacent channel interference requirements on a 802.11 system are based on a modulated signal and not a CW interferer. Further, given that the phase noise profile is likely to continue to roll off beyond a 12-MHz offset, the resultant interference with the desired signal would be less than that calculated above (the calculation above is considering worst-case interference at the edge of the band and for a CW signal). Therefore, in reality, the situation is much more complicated that that portrayed above. System level simulations are typically required to determine the exact impact of the out-of-band phase noise and ACI on PER. However, the reality of the situation, because of the reasons outlined above, would be less severe that the calculations above would suggest. It is important to note that achieving the above calculated phase noise performance (even for the worst case portrayed above) is fairly straightforward. One should note, however, that, in reality, today's WLAN systems are capable of sensitivities several decibels better than what the standard requires. Therefore, in order to achieve the state-of-the-art sensitivity levels, an additional 10 dB or so of margin would be required.

3.13   EFFECT OF PHASE NOISE ON OFDM SYSTEMS

In the case of the 802.11a/g OFDM signals, probably the most important impact of phase noise on the signal is the loss of orthogonality on the sub-carriers due to the inter-subcarrier interference caused by the PLL phase noise. Since each subcarrier has a bandwidth of 312.5 kHz, it is clear that the integrated phase noise of the closed-loop PLL across this bandwidth is quite important. The loop bandwidth of the 802.11 PLL is set by considera-tion of many criteria but is often selected to be in the few hundred kilohertz range. As such, since the phase noise typically rolls off after the loop band-width frequency, the interference of the subcarriers to the second and third adjacent subcarriers becomes less of an issue. The integrated phase noise over a 312.5-kHz bandwidth degrades the self-SNR of each subcarrier.

Another specification of interest for 802.11a/g systems is the integrated single-sided phase noise of the PLL across the entire channel (up to 8.125 MHz). This number is indicative of the PN-induced interference of any sub-carrier with the other subcarriers.

Beyond a 8.125-MHz offset, phase noise can cause interchannel interfer-ence (e.g., ACI).

In general, the effect of phase noise on the OFDM constellation (as compared to a single-carrier modulated signal) is "unconventional." As such, phase noise causes state spreading on the constellation points. Figure 3.28a shows the constellation diagram of an otherwise "ideal" 802.11a signal subject to severe in-band phase noise. Overall EVM is degraded to –26 dB. The EVM versus subcarriers for this signal is displayed in Figure 3.28b. Note that the phase noise impacts all subcarriers similarly, and it would be difficult to pinpoint phase noise as the reason for the signal impairment by solely looking at these plots.

There are certain digital techniques that may be used to reduce the impact of phase noise on the system performance. For example, a digital pilot-tracking-based phase noise cancellation scheme may be utilized. However, in general, these techniques are not very effective. For example, these techniques often cannot compensate for high offset frequency phase noise. Further these cancellation techniques operate on a post-FFT basis and cannot compensate for the inter-subcarrier "leakage" that occurs before the FFT operation. In general, the best solution is to have a high performance (low phase noise) radio PLL.

The details of the design of a low phase noise PLL are beyond the scope of this book. However, one should be aware that in order to maintain a low in-band (within the loop bandwidth of the PLL) phase noise response, components such as the crystal oscillator, frequency dividers, and charge pump should be designed with low noise levels. By going through the PLL equations, it can be shown that the noise contributions of these blocks will show up with a low pass characteristic as seen at the PLL output, with roll-off occurring at the loop bandwidth of the PLL. On the other hand, the VCO noise, for example, will have a high pass characteristic as seen at the PLL output and will therefore dominate the out-of-band characteristics of the PLL phase noise.

3.14   EFFECT OF FREQUENCY ERRORS ON OFDM

Another impairment affecting the WLAN system performance is the effect of frequency errors. The fundamental causes of frequency errors are crystal inaccuracies in the transmitter as well as the receiver side. As in any coherent modulation system, the OFDM receiver needs to track the transmitter frequency.

In a conventional single-carrier (non-OFDM) system, a frequency error would result in a continuous rotation of the constellation. However, in an OFDM system, a constant frequency error results in state spreading of the constellation.

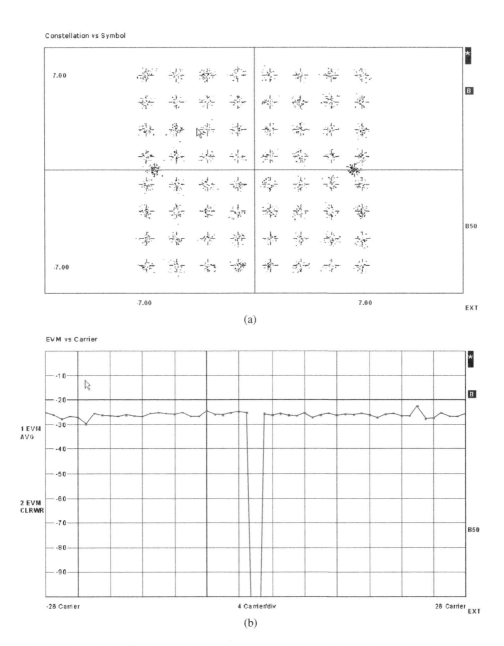

Figure 3.28 An 802.11a signal subject to severe in-band PLL phase noise: (**a**) constellation diagram; (**b**) EVM vs. subcarrier.

Recall that in an ideal system the OFDM subcarriers are orthogonal to one another over one symbol period of 3.2 μs. As an example consider the first and second subcarriers. By definition of orthogonality,

$$\int_0^{3.2\mu s} \sin(2\pi \times 312.5 \text{ kHz})t \, \sin(2\pi \times 625 \text{ kHz})t \, dt = 0$$

However, with a frequency offset Δf, orthogonality is lost:

$$\int_0^{3.2\mu s} \sin[2\pi(312.5 + \Delta f) \text{ kHz}]t \, \sin[2\pi(625 + \Delta f) \text{ kHz}]t \, dt \neq 0 \quad (3.4)$$

This results in state spreading as outlined above and a "smearing" of the constellation points.

The 802.11a standard requires that the transmitter have no more than 20 PPM of frequency offset. Since frequency accuracy is determined by the crystal, this typically translates into a specification that the crystal's frequency does not deviate from the required frequency by more than 20 PPM over parts, temperature, and time (aging). Certain crystal tuning and calibration techniques can be used, but this adds cost and some complexity. The

Table 3.1 Summary of Typical Analog Impairments, Impact on Single-Carrier Modulation Systems, and "Unconventional" Impact on OFDM-Modulated Systems

Impairment	Effect on Single-Carrier Constellation	Effect on OFDM Constellation	Method to Observe on VSA
I/Q Balance	Constellation skew	State spreading	Pilot tones on constellation diagram
Phase noise	Constellation arcing, especially at outer points	State spreading	LOGT-based measurement
Mismatched *I/Q* timing delays	State spreading	Constellation arcing, especially at outer points	EVM vs. frequency
Improper filtering		State spreading	EVM vs. frequency
Settling problems	State spreading	State spreading	EVM vs. time
Flicker noise		State spreading	EVM vs. frequency
Nonlinear distortion	State spreading	State spreading	CDF/CCDF
In-band spurs	State spreading	State spreading	EVM vs. frequency
Uncompensated Frequency errors	State spreading	State spreading	Spectrum plot

Note: A good method to identify the source of the analog impairment on the VSA is also listed.

cost and complexity have to be traded off against the use of a higher accuracy (but more expensive) crystal.

Frequency errors are typically corrected for in the receiver side in the digital domain. After the digital correction very little frequency error should exist in such a way that Equation 3.4 calculates to a number very close to zero. In the absence of any (typically factory) calibration and reasonably high accuracy crystals, there are certain advantages to correcting the frequency error in a "mixed-mode" fashion, especially in a direct-conversion system. This topic will be examined in more detail in the Chapter 5.

3.15   SUMMARY OF ANALOG/RF IMPAIRMENTS

Table 3.1 summarizes many of the impairments discussed in this section and lists some additional analog impairments impacting typical communications systems. In order to identify the unique behavior of OFDM-based systems, an additional column is added to identify the impact of the impairments on single-carrier systems. An additional column identifies a particular measurement to isolate the analog impairment on a vector signal analyzer.

CHAPTER **4**

Some Key Radio Building Blocks

So far we have discussed various system aspects of WLAN radio design. In doing so, we have discussed a variety of analog impairments which can impact the system performance. We will now transition to basic transistor level design of certain key building blocks used in various transceivers. This is, however, by no means a comprehensive examination of all building blocks used in a WLAN radio design. The interested reader is referred to the bibliography at the end of the book.

4.1 LOW NOISE AMPLIFIER

A LNA is a block that is typically placed as the first active component of a receive chain. As the name suggests, the primary purpose of an LNA is to provide some gain to the signal while adding minimal amount of noise. By doing so, it provides a larger signal at its output than its input, while the SNR at the output is only slightly worse than that at the input. The noise added by the circuit can be quantified by using the concept of NF as described earlier. Recall that NF can be described as 10 log(output SNR)/(input SNR).

Equivalently, NF can be described as the total input-referred noise of the circuit as compared to the noise from the source resistance:

$$\text{NF} = 10 \log\left(1 + \frac{N_i}{4kTR_s}\right)$$

Another important aspect of an LNA is that it is typically required to provide a known input impedance to the outside world. This is required because external passive (or sometimes active) components that connect to the LNA (filters, antennae, transmission lines, etc.) are designed with a certain characteristic impedance. If the LNA's input impedance deviates from the required characteristic impedance, the external passive components would not behave in the intended way. Most communications systems today are designed for 50 Ω characteristic impedance. This would require the LNA to ideally provide a 50 Ω real characteristic impedance and 0 Ω of imaginary (capacitive or inductive) impedance.

Further, to achieve a minimum possible NF given a certain external source resistance, a certain optimal LNA input impedance would be required. This input resistance is not the same (but often not too vastly different) as what would be required for an optimal power match. One important goal of the LNA designer is to balance these input matching needs.

Let us assume that an input 50-Ω real match is required. There are several ways of achieving real 50 Ω input impedance on the LNA. In the simplest approach (shown in Fig. 4.1), a 50-Ω shunt resistor can be placed at the input of the LNA (while any input capacitance would be tuned out with an inductor and presumed to be high impedance at resonance). Unfortunately, by utilizing such a scheme, even with a noise-free LNA, the NF of the LNA will be limited to a minimum of 3 dB since $R_i = R_s$:

$$\text{NF} = 10 \, \log\!\left(1 + \frac{N_i}{4kTR_s}\right)$$

$$= 10 \, \log\!\left(1 + \frac{4kTR_i + \cdots}{4kTR_s}\right)$$

$$> 3$$

In reality, the NF will be larger than 3 dB with this matching scheme. Clearly this is not a desirable scheme for systems that require optimal sensitivity.

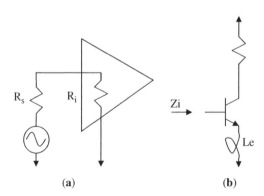

(a)   (b)

Figure 4.1 (a) Input matching using "brute-force" resistive termination. (b) Narrowband input matching using inductive degeneration in common-emitter (or common-source) configuration.

An alternative scheme which is quite popular for *narrowband*[72] applications is the use of inductive emitter degeneration for bipolar and source degeneration for CMOS transistors. In this scheme, an inductor is placed at the emitter terminal of the transistor (Fig. 4.1b). Because of the capacitive roll-off nature of the AC gain (β) of the device, when this inductor is reflected back to the input on the base terminal, it is transformed to a resistive component:

$$Z_i = r_b + (\beta + 1)j\omega L_e - j \ldots$$

$$= r_b + \frac{\omega_T}{j\omega} j\omega L_e + j\omega L_e - j \ldots$$

$$= r_b + \omega_T L_e - j \ldots$$

Here, r_b is the base resistance of the device, ω_T is the unity gain frequency of the device, β is the AC gain of the device, and L_e is the emitter degeneration inductance. As can be seen, the term $\omega_T L_e$ is a purely real component. The remainder capacitance seen at the base can easily be canceled by using an inductor in the base side. In this approach, since Re(Z_i) is set purely by the reactive component of L_e (ignoring r_b), no noise penalty is incurred.

Similarly, in a CMOS approach, r_b is replaced with r_g (gate resistance) which would represent the geometric-based polysilicon gate resistance as well as the non-quasi-static noise due to the MOSFET channel resistance. Otherwise the same equation applies.

It is important to note that a certain device in a certain technology can obtain an optimal NF for a unique input impedance. This input impedance is often not the same impedance as the optimal power match (typically 50 Ω). The LNA designer, therefore, often designs the circuit for an optimal noise match which simultaneously satisfies the minimum required input impedance match (S11).

Using this scheme, LNA noise figures of 1 to 2 dB are achievable in standard CMOS and SiGe processes at the 2.4- and 5-GHz bands.

If the LNA output is brought off chip (e.g., for image signal or noise filtering) output impedance matching will also be required. For WLAN applications this is typically avoided, and image and noise filtering is achieved by a variety of other means depending on the receiver architecture.

Another important characteristic of an LNA is its linearity. As described earlier, often an LNA is exposed to very large signal interferers with mini-

---

[72]In this context a bandwidth of 1 GHz across the 5-GHz band for the 802.11a band is considered to be narrowband.

mal filtering support. If a large incoming signal is due to the desired signal being large, gain control in the LNA can be utilized to reduce the amplitude of the signal at the output of the LNA (reasonably high linearity at the input of the LNA is still required). However, if the LNA is exposed to a large interfering signal while receiving a small desired signal, it would need to have good linearity as well as good noise performance at the same time. Achieving high linearity performance in the LNA while maintaining a low NF is quite challenging.

Figure 4.2a shows the schematic of a more realistic LNA used in a 5-GHz WLAN transceiver (Bouras et al., 2003). The L_s devices are used to obtain the proper real part of input impedance and the L_{gs} devices are used to cancel out the capacitance seen at the gates. The L_{ds} devices are used to resonate out all the capacitance existent on the output section. This includes any capacitance due to the LNA output devices as well as the input stage of the following mixer.

In this scheme a differential LNA is used as the first active component of the receive chain. The choice of a differential approach is quite popular with many high integration systems, since a certain amount of immunity is obtained to any common mode interference that may be present at the chip–package interface[73] or on the chip substrate. The penalty for using a differential design is that extra current would have to be consumed in order to achieve the same NF as the equivalent single-ended design.

The design shown in Figure 4.2a also incorporates a differential cascode design. The use of cascodes is also quite popular since they provide many advantages. For example, the impact of Miller capacitance is reduced and therefore a higher gain can be achieved from a single stage. Further, a significant improvement to the reverse isolation of the amplifier (S12) is achieved. This simplifies the input and output matching or tuning of the LNA by making the matching networks relatively independent of one another (a "unilateral" circuit). Due to the higher isolation, the cascodes also help reduce the possibility of oscillations in the circuit.

At low frequencies, the cascodes do not contribute to the NF of the LNA. As the frequency is increased, however, the cascodes will contribute to the LNA NF since the output impedance of the input transconductor devices (at their drains) are reduced with increased frequency (due to the parasitic capacitance at these nodes). In order to minimize the impact of the cascodes on NF, the parasitic capacitances at the drain of the input de-

---

[73]The chip–package interface is a very common point for the ingress of noise or interference on to the small received signal. The ingress is often due to magnetic couplings due to bond wires and/or traces on the package.

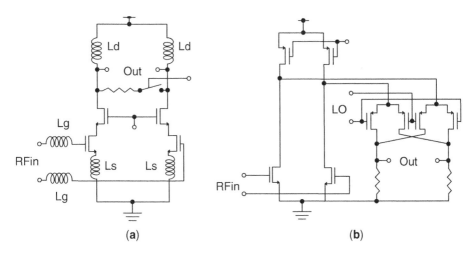

(a)   (b)

Figure 4.2 (a) A more realistic WLAN LNA configuration. (b) A WLAN receiver folded mixer cell.

vices must be minimized by utilizing proper design and layout techniques.[74]

The LNA of Figure 4.2a utilized switched-resistance gain control at its output stage. By switching in a variety of resistors, the output tank is dequeued, and as a result the gain of the LNA is reduced. In designing such a gain control scheme, the designer should size the switch properly:

- If the switch is sized too small, the on resistance of the switch can become comparable to that of the resistor and as a result the desired reduction in gain may not be obtained. Even if the resistance of the switch is accounted for in the design, there would be a significant part-to-part and over-temperature gain variation due to the variability of the resistance of a MOS switch over these parameters. Further, the nonlinear resistance of the switch may limit the linearity performance of the LNA when the switch is on.

- If the switch is sized too large, its parasitic capacitance can be significant and the effect of the resistor dequeuing may be observed even when the switch is off. This will result in a reduction in the maximum achievable gain of the stage and a reduction in the range

[74]Layout techniques include the removal of the diffusion-to-metal contacts at these nodes and sharing the diffusion implant region between the source of the cascode device with the drain of the transconductor. Circuit techniques include the use of inductors at these nodes to tune out the capacitance (at the expense of added area).

of gain control that is achieved. Further, the nonlinear capacitance of the switch may limit the linearity of the LNA, especially when the switch is off.

In reality, quite often the switch network is implemented in a fully differential fashion to preserve the symmetry and common-mode capabilities of the circuit. As such the switch–R network would be implemented as an R–SW–R or an SW–R–SW configuration.

Many other methods can be used to implement gain control in the LNA. For example, the cascode and transconductance devices can be grouped with the total width of each group set by the gain step required. Then the cascode gates of each group can be controlled individually to obtain the desired level of signal attenuation. In this approach, the gate voltage of each group is tied to the proper cascode voltage if the section is to be turned on, and it is grounded if the stage is to be turned off. In this scheme, effectively the input transconductance is reduced to reduce the block gain. An advantage of this scheme is that the DC is reduced at higher attenuation settings. On the other hand, as the attenuation setting is increased, the input impedance and therefore the S11 of the block are modified.

Modifications of this general scheme have also been utilized. For example, instead of shutting down the cascode device of a particular section, the output current of that section can be diverted away from the load, thereby reducing the effective transconductance delivered to the load and reducing the gain. In this scheme the S11 of the block as a function of the gain setting remains relatively constant.

The LNA shown in Figure 4.2a and described above is implemented in 0.18-μm CMOS technology and achieves a maximum gain of 18/10 dB in the high gain/low gain modes. In the high gain mode, this particular implementation achieves an LNA NF of 3 dB for the lower 802.11a bands and a NF of 4.5 dB for the highest band. The overall system NF of this implementation is ~6 dB.

Current state-of-the-art CMOS LNAs are capable of obtaining <2 dB NF in the 802.11a bands. Systems utilizing these latest LNAs are capable of achieving <4 dB NF in the 802.11a band.

4.2   MIXER AND ITS LOCAL OSCILLATOR BUFFERS

Another key building block in the design of wireless transceivers is the frequency conversion block or the mixer. The symbol for a mixer is simply a multiplier. If two sinusoids f_1 and f_2 are applied to a multiplier, frequency conversion is achieved at the sum $(f_1 + f_2)$ as well as the difference $(f_1 - f_2)$ of the two frequencies:

$$\cos(f_1)\cos(f_2) = \tfrac{1}{2}[\cos(f_1 + f_2) + \cos(f_1 - f_2)]$$

Many schemes can be used to obtain frequency translation. One very popular scheme used to obtain frequency conversion is a "Gilbert quad" to obtain current-mode multiplication (Fig. 4.3). In the case of a bipolar Gilbert quad with the inputs

$$I_{in,d} = A_I \cos(\omega_{RF}t) \qquad V_{LO} = A_{LO}\cos(\omega_{LO}t)$$

one can get

$$I_{out,d} = I_{in,d}\tanh\left(\frac{V_{LO}}{2V_T}\right)$$

To maximize the conversion gain of the mixer and also improve its linearity and noise performance, it is best to operate the Gilbert quad in a fully switched mode rather than as a multiplier. In such a scheme, if $A_{LO} \gg 4V_T$ (V_T is the thermal voltage and is equal to 26 mV at room temperature), the input current is "fully" switched. This is equivalent to multiplication by a unit square wave with a fundamental component amplitude $4/\pi$:

$$I_{out,d} = A_I\frac{2}{\pi}\cos[(\omega_{RF} \pm \omega_{LO})t]$$

Similar results can be obtained for a CMOS mixer. However, in order to ensure full switching, the amplitude of the LO has to be significantly larger than that required in a bipolar mixer. This is because of the exponential relationship between I_c and V_{be} in a bipolar transistor as compared to the square-

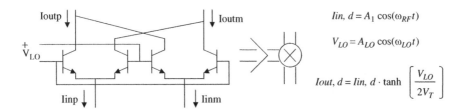

Figure 4.3 Basic Gilbert cell mixing quad utilizing bipolar transistors. Associated basic equations are also shown for when the quad is used in a multiplier mode.

law (or linear in the case of short channel length) relationship between I_d and V_{gs} in a MOSFET. As a result, the LO buffers used in CMOS designs typically burn significantly more current than their bipolar counterparts.

The schematic for a receiver mixer cell is shown in Figure 4.2b (Bouras et al., 2003). This mixer is utilized in an 802.11a direct-conversion receiver. This mixer is implemented in a fully differential doubly balanced fashion to maintain a high common-mode rejection ratio and also provide a high degree of RF and LO rejection at the baseband output. The input transconductance is made of NMOS devices for maximum g_m and minimum NF. Instead of feeding the resultant output current from the transconductor devices directly into NMOS Gilbert quads, the output current is folded using the top PMOS current sources and diverted into the PMOS Gilbert quad (this structure is known as a folded cascode in opamp design). Theoretically the use of PMOS quad results in a reduced flicker noise contribution to the mixer NF due to the switching devices.[75] In reality, however, the use of a PMOS quad would require the designer to use larger size devices for these switches (as compared to the NMOS devices) for the same conversion gain performance. For a given LO buffer power consumption, this would result in a reduced swing at the LO port. As a result, a longer period of time would be spent at or around the zero-crossing point of the quad devices, resulting in a higher contribution by these devices to the mixer noise. Further, the thermal noise of the PMOS current-source devices can also contribute significantly to the overall NF. The optimal overall design, therefore, would depend on the particulars of the device technology and design techniques used. It is also important to note that, due to reduced power supply voltages available on modern technologies, a folding technique may be required to accommodate the voltage headroom constraints of the design.

This particular design achieves a single-sideband NF of 12 dB and an IIP3 of +4 dBm while operating at the 5-GHz band. The mixer is typically the limiting block in the receiver linearity performance. In order to be able to receive large input signals, it is therefore desirable to achieve a higher IIP3 that that achieved by this design. Current state-of-the art active mixer designs utilized for 802.11a can achieve IIP3s of ~15 dBm. Passive mixer designs can achieve IIP3s of > 20 dBm while obtaining NFs of < 10 dB.

---

[75]The fundamental reason that the flicker noise of PMOS and NMOS devices is different is because the underlying phenomena causing the flicker noise are believed to be different. In the NMOS device, flicker noise is believed to be due to the random trapping and detrapping of the free carriers in the channel and the resultant change in the number of free carriers available in the channel. For a PMOS device, on the other hand, the cause of flicker noise is believed to be the bulk mobility fluctuations through scattering phenomena. The exact cause of flicker noise is an active device physics research area.

One method to linearize the mixer G_m is the use of the scheme shown in Figure 4.4 (Bouras et al., 2003). In this scheme, the input signals are applied at the positive terminals of the opamps U1 and U2, rather than directly at the inputs of the PMOS transconductors M1 and M2. The opamps are configured as unity gain buffers and apply the differential voltage across the variable-switched degeneration resistor R_x. The output current is then set by the differential voltage V_{ind} divided by the degeneration resistance R_x. Effectively, the opamps U1, U2, the transistors M1, M2, and the variable resistor Rx form the composite differential transconductor for this circuit. The loop gains of the opamps allow for this composite transconductor to have a very high linearity. In this implementation, an IIP3 of +20 dBm (for a hypothetical 50-Ω reference) is achieved. An example of a mixer, in this case a transmit-side variable-gain mixer utilizing this transconductor linearization scheme, is shown in Figure 4.4 (Bouras et al., 2003). As explained earlier, it is desirable to maintain a large amplitude on the baseband section of a transmitter in order to reduce the LOFT due to baseband DC offsets. However, in order to maintain a large baseband signal, the baseband chain would need to have a high degree of linearity. In particular, the transconductance stage of the up-converting mixer needs to be linearized in some fashion.

The gain of this stage is programmable and can be adjusted by switching in the proper value of resistance R_x. In this implementation 27 dB of programmable attenuation is achieved. It is important to note that maintaining the desired LOFT at the higher attenuation settings is more difficult since the signal level is attenuated but the DC offsets may not be. Therefore the

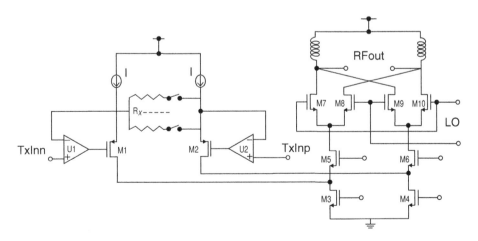

Figure 4.4 Highly linear variable-gain up-conversion mixer cell utilized in some WLAN applications. The loop gain of the opamps is utilized to maintain high linearity in this cell.

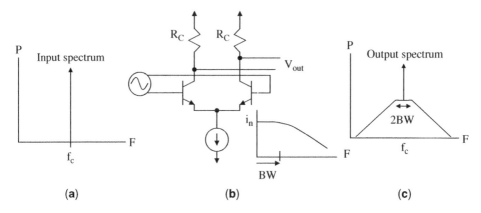

Figure 4.5 (a) Ideal hypothetical LO signal applied to LO buffer. (b) Basic bipolar LO buffer showing noise on tail current source. (c) Output spectrum at buffer output showing up conversion or tail current source and corruption of desired signal.

LOFT in decibels relative to the carrier will degrade at the higher attenuation settings.

The output current of the transconductor stage is then folded through the folding NMOS current sources and cascode devices M3 to M6 and fed into the Gilbert quad M7 to M10. The signal is then unconverted directly to the desired RF in the direct up-conversion transmitter. Note that in this case the flicker noise associated with the switching quad is not of concern, since, unlike a down-conversion mixer, their flicker noise will not show up at the output spectrum. At the same time the flicker noise of the input transconductor devices, the current sources, and so on, do need to be taken into account. The inductors at the output are used to tune out the parasitic capacitances at the output of the mixer.

In the actual implementation, two baseband paths and mixers are used to accommodate the quadrature image-reject up conversion of the baseband signal. The outputs of the mixers are connected to one another, and a shared pair of inductors is used as the common load.

Another important but often overlooked radio block is the LO buffer(s) used for applying the required synthesized LO signal to the receive and transmit mixers. The VCO is rarely directly coupled to the mixers even if the VCO frequency is the same as the LO frequency. This is because it is important to maintain a high Q tank on the VCO in order to obtain a low phase noise. Routing lines used between blocks often have a significant amount of resistance associated with their capacitance (and inductance). As a result, they could dequeue the VCO tank and degrade phase noise. The longer the routing distance is, the more the potential for dequeuing the tank would be. By adding a local buffer close to the VCO, the VCO can be isolated from floor planning and routing issues, and the modeling of the VCO parasitics are also simplified.

Another reason to use buffers between the receive and transmit mixers and the VCO is to provide reverse isolation. In a half-duplex system such as that of 802.11, as the system switches between receive and transmit, the load seen by the driving circuit can change. This load change can cause load pulling in the VCO. The more reverse isolation provided by the LO buffers, the less the chance of VCO load pulling.

The primary task of the LO buffers are to provide the proper amplitude LO signal to the mixers without adding significant noise to the synthesizer-generated signal. For bipolar mixers, this amplitude can be fairly small (a few times V_T), but for MOS mixers it needs to be significantly larger than that (often a few times $V_{gt} = V_{gs} - V_t$). In order to maintain the required amplitude over process, temperature, and supply, it is often desirable to hard switch the buffer's devices (Fig. 4.5). However, with the buffer hard switched, the noise present at the tail current source will up-convert (with an up-conversion gain of $2/\pi$) and mix with the desired LO signal. In a sense the buffer acts as an up-converter mixer to the tail current noise. It is therefore quite important to maintain low noise on the bias lines. Bias line noise (on the VCO or the buffers) is often one source of phase noise seen on measured response that may have been missed in simulations. This is because, during simulations, in order to speed up simulations or resolve convergence problems, designers may not include the entire bias generation circuitry. If a designer opts not to include such biasing blocks in PN simulations, it is advisable to use behavioral biasing models that include the noise sources. Figure 4.5 displays how the baseband bias noise can up convert and corrupt the desired ideal LO signal.

Depending on the frequency of operation and devices used, LO buffers can be resistively loaded or inductively loaded. For WLAN applications, especially at the lower 2.4-GHz band, SiGe bipolar buffers can often be designed with resistive loads to achieve a small area. At the higher frequencies and especially for CMOS designs where large parasitics are present and large swings are required, inductively loaded stages are often used to allow the resonating of the parasitic capacitances. In general, there is a trade-off between the area consumption of the buffer and its power consumption.

Especially in CMOS designs where large LO swings are required, the design of the LOGEN circuitry and buffers is a very important part of the overall design and floor planning. Often LOGEN should be included in the very early stages of the top level floor planning. Also often LOGEN consumes a significant percentage of the power consumption of the chip and should therefore be budgeted for accordingly. This is especially true if low noise and high swing performance is required out of the LO buffers.

4.3 POWER AMPLIFIER

Another major building block used in various transceivers is the power amplifier (PA). The term *PA* is probably one of the most liberally used in the industry. Depending on the context, the system and the person, a PA can produce –5 dBm of output power or it can generate +40 dBm of output power! Further, the term *output power* itself should be elaborated on. Sometimes the specified power level is for the "saturated" output power of the PA (the maximum possible output power of the PA, where increasing the input signal would generate no further increase in the output power). At other times, the specified output power level is for the 1-dB compression point of the PA. Yet at other times, the specified power level is the maximum "linear" output power level of the PA. This latter specification requires further clarification as to what modulation type is being transmitted by the PA. It should be evident that designing a PA may be quite challenging or rather simple depending on the context.

As discussed earlier, the PAR of a 802.11 OFDM signal can be as large as 17 dB. The transmitter blocks, in particular the PA, is therefore required to produce substantially larger peak signals than the average signal. In order to maintain linearity, this typically requires a significant "backoff" from the 1-dB compression point of the PA (i.e., the average transmit power is set to be several decibels below the P1dB point of the PA). This fact significantly reduces the maximum available efficiency obtainable from a 802.11a/g PA. Fortunately, in reality, the largest peaks of the OFDM signal occur very infrequently, and therefore a much smaller backoff produces reasonable system performance while increasing the attainable efficiency.

An example of a CMOS three-stage power amplifier implemented for 5-GHz WLAN applications is shown in Figure 4.6 (Zargari et al., 2002). In order to maintain linearity, the PA is operated in class A mode. All stages are cascoded and operate off of a 3.3-V supply. The cascode devices allow for better stability, independent input and output matching, and a better device reliability in the presence of large signals present at the drains of the output devices. Each stage is capacitively level shifted so that the optimal common-mode voltage for each stage can be properly set. Capacitors are implemented using stacked metal 2 through metal 5 layers. Inductive loads are used to tune out the large capacitance present at the output of each stage and the input of the following stage. The overall output of the PA is a differen-

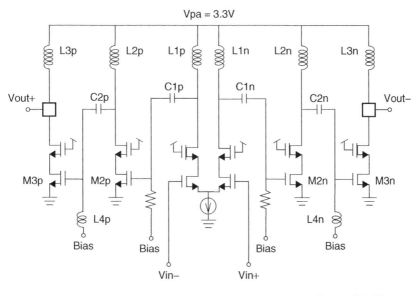

Figure 4.6 A three-stage CMOS PA achieving a saturated power of +22 dBm in 0.18-μm technology and utilizing a 3.3-V supply.

tial signal which requires an external balun to convert into a single-ended output. This PA achieves a saturated output power of 22 dBm and a maximum transmit power (chip referred) of ~12.5 dBm for a 54-Mbps OFDM-modulated signal. This PA is of interest as one of the first published PAs for WLAN. However, the achieved efficiency is low since no special techniques are utilized to enhance the efficiency of this PA.

Power amplifiers used in WLAN applications are one of the most challenging to design. PAs and diversity and time-duplexing switches are the only main transceiver building blocks that are not commonly integrated along with the rest of the transceiver blocks in WLANs. There have been published examples of these functions being integrated on the same CMOS substrate as the rest of the system. The achievable performance, however, is often quite inferior to that obtainable by using other process technologies (SiGe and GaAs).

When it comes to design of power amplifiers, CMOS devices in general have many shortcomings as compared to GaAs (and even SiGe) devices. A unit of merit that is quite applicable to PA design is the unity gain current frequency and breakdown voltage product of the devices. Although state-of-the-art CMOS devices have very high f_T's, they suffer from very low break-down voltages. GaAs devices (both PHEMT and HBT), on the other hand, simultaneously offer very high f_T (and f_{max}) and very high breakdown voltages. Further, in order to deliver a certain output power, the output devices have to be sized properly and need to operate at the proper current density. As a result, for a given output power, a CMOS device needs to be sized much larger than the equivalent SiGe GaAs HBT device, resulting in much larger parasitic capacitances. This is one of the factors causing a reduction in the maximum attainable efficiency from the CMOS PA. In addition, this factor reduces the maximum available gain per stage out of CMOS PAs.

Another important shortcoming in CMOS processes is the lack of a *through-wafer via*. A through-wafer via is commonly available on GaAs processes and allows for a very low inductance ground (~50 to 100 pH) to be placed anywhere on the die. The lack of such a via technology in CMOS processes essentially eliminates the possibility of designing a single-ended CMOS PA at higher frequencies. This is because a small ground inductance on the source of the PA device significantly reduces the gain attainable from the CMOS gain stage. Clearly, the higher frequency, the more of an issue this becomes, as the magnitude of the "inductive degeneration" rises with increasing frequency. Because of this problem, reported CMOS PAs have been implemented in a differential fashion. As such the virtual ground at the common-source point of the transconductor devices minimizes the impact of the lack of a low impedance ground. From a small-signal-gain point of view, the use of differential circuitry essentially eliminates this problem. However, from a large-signal point of view, the second-order harmonics of the input signal still require a low impedance ground at the common-source point (since they are essentially a common-mode signal rather than a differential signal, the common-source point is not a virtual ground to them). Several techniques have been proposed in the literature to work around this problem.

As mentioned earlier, the gain of a CMOS PA, especially at high frequencies, could be quite limited. As such, for a given output power, the input of such a PA would observe a larger signal swing as compared to a PA with a high gain. The linearity of the PA can therefore be limited by the linearity of its transconductance stage. In order to generate a large linear signal out of the transmitter while maintaining reasonable power efficiency, a class AB power amplifier driver and power amplifier stages can be utilized (Fig. 4.7). The transconductance of the class AB stages, however, need to be quite linear in order to maintain the linearity required for 64-QAM OFDM modulation. Further the transmitter stages need to have reasonable amount of gain. This rules out source (inductive or resistive) degeneration as a possible solution.

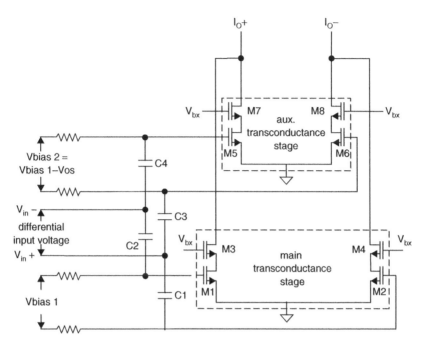

Figure 4.7 Highly linear transconductor stage utilized in CMOS power amplifier for WLAN.

Finally, it is desirable to allow for maximum voltage swing on the stage. A pseudodifferential grounded differential amplifier ("main transconductance" stage of Fig. 4.7) has minimal headroom requirements. However, as shown in the plot of transconductance versus incoming voltage of Figure 4.8, such a stage exhibits large variation in transconductance as a function of the incoming voltage. By utilizing an additional grounded pseudodifferential amplifier "auxiliary transconductance stage" operating with the same incoming AC signal but shifted operating points implemented through AC-coupled stages of Figure 4.7 (Behzad et al., 2004a), additional transconductance gm_sub of Figure 4.8 can be generated. When these transconductances are added at the drains of the differential pair transistor cascodes, a much more linear transconductance curve (Gm_total of Fig. 4.8) would be generated. A significant reduction in transconductance variation has been achieved. This results in significant HD3 and IM3 reduction, an increase in the 1-dB compression point, and a higher linear maximum power operating point. The improvement in HD3 or IM3 performance for this linearization scheme can be solved mathematically; however, the deviation is fairly algebraically involved and will not be presented here. Several points are in order:

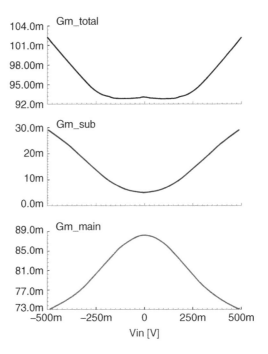

Figure 4.8 Plots of DC transconductance for main stage, auxiliary stage, and overall stage (linearized).

- The technique can be easily applied to 2 to N stages. In this example, a two-stage technique has been shown. Trade-offs can be made between maximum peak-to-valley deviation of the transconductance, the "in-band" ripple of the transconductance curve, and the linearized range of the transconductance curve. Due to the complex shape of the overall transconductance curve, simple extrapolation concepts such as intercept points no longer apply.

- The differential pairs described in this linearization technique operate in the class AB mode with grounded sources. The operation in this mode is quite different than linearization techniques operating in the class A mode with a constant tail current source.

- Unlike many other linearization techniques, such as emitter degeneration, this technique can achieve fairly large *and* linear transconductance levels simultaneously.

- For a given input voltage swing and a particular number of stages with certain device sizes, an optimal offset voltage can be found which results in a very large improvement over an uncompensated G_m stage. However, if properly designed, the scheme is tolerant of reasonable size device mismatches and resultant offsets and can result in several decibels in linearity improvement even in the presence of 3-sigma device mismatches.

As will be seen shortly, the general technique is applicable to low frequency applications also.

It should now be clear that traditional design techniques utilized in the design of SiGe and GaAs PAs need to be augmented with newer and more innovative techniques in order to obtain reasonable performance out of CMOS PAs. These techniques may include various analog, digital, and mixed-mode linearization techniques, dynamic bias boosting techniques, envelope tracking techniques, digital predistortion techniques, and even some of the most exotic techniques such as LINC (linear amplification using nonlinear components). Needless to say, CMOS PAs are a very active research area.

4.4 FULLY INTEGRATED VCO

In recent years, fully integrated VCOs with adequately high performance for general radio needs have become commonplace. Similarly, all of today's commercial WLAN radios utilize integrated VCOs.

In a VCO, a resonant structure (typically an *LC* tank) sets the center frequency of oscillation and a control voltage can then adjust the frequency of operation by changing the tank resonance frequency. Due to parasitic losses in the tank, an active circuit needs to provide gain and compensate the losses in the tank in order to be able to sustain an oscillation. A cross-coupled pair of active devices (CMOS or bipolar) can provide a *negative resistance* through the use of positive feedback. This negative resistance acts to cancel the positive resistance present in the tank due to the losses. As a result a sustained oscillation condition is achieved.

Some fundamental equations govern the basic operation of most integrated LC-based oscillators. The center frequency is set by

$$\omega_0 = \frac{1}{\sqrt{LC}}$$

where ω_0 is the frequency of oscillation, L is the effective inductance of the tank, and C is the effective capacitance of the tank. Note that all parasitic capacitances and inductances must be considered in applying the above equation. The effective parallel resistance (R_p) of the tank can be calculated by

$$R_P = \omega_0 L Q_L$$

where Q_L is the Q of the inductor used in the tank. For a VCO whose bias current (I_{tail}) is set by a tail current source, the maximum voltage swing (V_{max}) is given by

$$V_{max} \approx I_{tail} R_P$$

In order to ensure that oscillation starts, the active devices need to have enough loop gain to compensate for the losses in the tank. Some margin is also necessary here. A reasonable rule of thumb is to design the VCO such that the large signal transconductance (G_m) satisfies the relation

$$G_m > \frac{3}{R_P}$$

A very important characteristic of a VCO is its phase noise. Phase noise is often specified in decibels relative to the carrier per hertz at a certain frequency offset from the center frequency. Phase noise in oscillators has been the subject of intense research for the past few decades and continues to attract significant attention from researchers. The following relation does a good job of relating phase noise to physical characteristics of the VCO:

$$PN(\Delta\omega) = \frac{\Gamma_{rms}^2 i_n^2}{2 I_{tail}^2 Q^2} \left(\frac{\omega}{\Delta\omega} \right)$$

where $\Delta\omega$ represents the offset frequency of interest, ω is the center frequency of the VCO, I_{tail} is the tail current, Q is the Q of the tank, and Γ_{rms} is the RMS value of the impulse sensitivity function for the VCO and is related to how the timing of the injection of the noise affects the zero-crossing point of the VCO and therefore its phase noise.

This equation shows that, in order to minimize phase noise, the voltage swing of the VCO should be maximized. This requires a large tail current. However, care must be taken to ensure low levels of noise on the bias lines. The bias lines are often the source of degraded PN performance in a VCO. It is common to heavily filter the tail current sources to ensure low levels of noise on the bias lines.

At the same time, the noise added by the switching transistors near the zero-crossing point is proportional to their transconductance. It is therefore desirable to keep G_m just large enough to ensure oscillation startup. From this point of view, a MOSFET device is desirable to a bipolar device since the MOSFET has a smaller transconductance for a given current. Unfortunately, however, as compared to a bipolar device, a MOSFET de-

vice also has very large flicker noise. This flicker noise is typically unconverted to the RF and causes significant degradation in close-in PN of the
VCO.

So to summarize, for low PN, a high Q tank (which usually requires large
area), large bias current (therefore high power consumption), and low bias
current noise are required. The bias current noise is required to be low at
baseband frequencies, the fundamental frequency of operation of the VCO,
as well as at the harmonics of the VCO frequency. This is because the noise
at these various harmonic frequencies can convert in frequency (due to the
nonlinear nature of the operation of the VCO) and show up at the fundamental frequency and degrade PN.

A fairly common topology used in the design of CMOS VCOs is shown
in Figure 4.9 (Bouras et al., 2003). Several interesting points can be made
about this VCO topology:

1. Two pairs of cross-coupled active devices are used to generate the required negative G_m: an NMOS pair and PMOS pair.
2. In theory, the use of these two pairs of devices creates more symmetric rise and fall times. This symmetry helps reduce the up conversion
   of flicker noise from baseband to the operating frequency of the VCO.

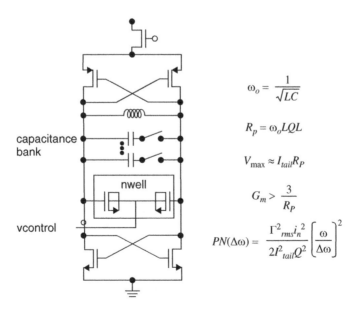

$$\omega_o = \frac{1}{\sqrt{LC}}$$

$$R_p = \omega_o LQL$$

$$V_{max} \approx I_{tail} R_P$$

$$G_m > \frac{3}{R_P}$$

$$PN(\Delta\omega) = \frac{\Gamma^2_{rms} i^2_n}{2 I^2_{tail} Q^2} \left[\frac{\omega}{\Delta\omega}\right]^2$$

Figure 4.9 Example of fully integrated CMOS VCO utilized in some WLAN radios. Some
common VCO design equations are also listed.

3. The use of the top and bottom pair of cross-coupled devices reduces the swing of the output signal of the VCO. This may result in a higher reliability design as the devices are not stressed beyond their "normal" region of operation. However, at the same time, a reduced swing output would degrade the PN performance of the VCO. Further this design requires a larger voltage headroom than some alternative designs.

Many applications require a wide tuning range out of the VCO in order to accommodate tuning to a variety of channels. For example, in a worldwide 802.11a implementation, the LO is required to tune from 4.9 to 5.9 GHz. Given the very low supply voltages available in modern process technologies, this wide tuning range would require a very high gain VCO. For example, if the compliance range of the charge pump is 1 V, a VCO gain of 1 GHz/V would be required to cover the 802.11a band (in other words a 1-μV change on the control line would change a 1-kHz shift in the center frequency of the VCO!). The control voltage of the VCO is one of the most sensitive nodes of a transceiver, and this degree of sensitivity would certainly create PN problems in the system (due to noise directly coupling on the control line or through the supplies). It is therefore highly desirable to reduce the gain of the VCO. One approach would be to introduce banks of switched capacitors in the VCO that would allow discrete step adjustments to the center frequency. With this scheme, the switched capacitors can be used to place the VCO close to the desired frequency, and the continuous-voltage-control line would then be controlled by the PLL and adjust the varactor capacitance to lock the exact frequency. Often automatic calibration schemes are used to find the required number of switch capacitors to be turned on. The continuous control would have to have enough range to be able to maintain a lock over the temperature range of operation. Otherwise, periodic partial switched-capacitor calibrations would be required to maintain a lock over temperature. Using this scheme a K_{vco} of as low as 20 MHz can be used to cover the entire 802.11a band. Care must be exercised to ensure that the queue of the switched-capacitor network is high in order to avoid dequeuing the tank. This scheme has been deployed in the design of Figure 4.9 and is probably the most common scheme utilized in VCOs to enable wide tuning range while maintaining good PN performance.

Alternatively, multiple VCOs can be utilized to cover wide bandwidths. This approach often achieves a lower power consumption and better PN, at the expense of a larger area. Yet as another alternative, switched-inductor networks can be utilized in conjunction with the switched-capacitor network. Care should be taken to ensure than the switches utilized with the inductors do not dequeue the tank significantly as this will result in a degraded PN performance.

4.5 MULTIFREQUENCY (STACKED) MIXER

Another radio building block that is useful in designing certain type of transceivers is a stacked (multifrequency) mixer as shown in Figure 4.10 (Behzad et al., 2003). This circuit provides two consecutive mixing operations in one circuit block and can be useful in certain receivers, transmitters, as well as LO generation circuitry. One example of the utilization of such a circuit will be provided in the case study design of Chapter 6. In this application, an LO is to be synthesized in a direct conversion transceiver. In order to avoid any pulling issues, the VCO is operated at two-thirds the RF. Further a low frequency correction signal, F_{AFC}, is needed to correct for the crystal oscillator frequency inaccuracies. The objective is to generate an LO signal at

$$F_{l0} = F_{afc} + 1.5F_{vco}$$

Two consecutive mixing operations can clearly be performed with two standard Gilbert-style mixers. However, by instead utilizing the circuit

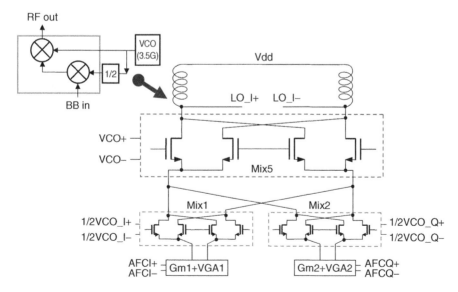

Figure 4.10 Multifrequency stacked mixer implementation used in generating $F_{rf} = 1.5$ $F_{vco} + F_{afc}$ in LOGEN block of WLAN radio.

shown in Figure 4.10, one additional stage of current-to-voltage conversion and then voltage-to-current conversion are eliminated. As a result, one set of load inductors as well as one set of transconductors can be eliminated, resulting in a smaller die area, lower power consumption, and higher linearity. In this application of this design, the baseband input voltages are applied at the transconductor inputs (labeled G_m + VGA at the bottom of the schematic). The signal is then converted to a current and applied to an optional current-mode variable-gain block. The gain controlled output is then passed through the first set of mixers (Mix 1, Mix 2) and unconverted to 0.5 times the VCO frequency. The output current is then further applied to mix 5 so that it can be mixed with the LO frequency of this mixer running at the VCO frequency. The overall unconverted current at 1.5 times the VCO frequency is converted into a voltage at the output of the mixer. Signals at $1.5F_{vco} \pm F_{afc}$ are therefore generated at the output of this mixer.

The basic idea of a stacked mixer is also applicable to the design of superheterodyne receivers and transmitters if no other signal processing (e.g., gain, filtering) is required at the intermediate stage.

It is important to note that the headroom constraints of this design have to be carefully analyzed due to the existence of multiple stacked devices.

4.6 OPEN-LOOP TRANSCONDUCTANCE LINEARIZATION CIRCUIT

Various techniques can be utilized to achieve high linearity in circuit building blocks. Feedback techniques provide many advantages but for certain applications may not be the optimal solutions. For example, they may require a relatively large area to compensate the loop or they may consume more power than an open-loop technique. Further, feedback techniques at high frequencies are not as effective since obtaining high amounts of loop gain at these frequencies is not possible. A generic open-loop transconductance linearization technique is demonstrated in Figure 4.11 (Behzad et al., 2003). This general technique in slightly varied forms is applicable to receivers, transmitters (see, e.g., Fig. 4.7), and LO generation circuitry. In this example the technique is applied to a LO generation path and will be discussed in some detail in Chapter 6. The objective is to generate a highly linear and reasonably large transconductance output from this block while consuming a minimal amount of supply headroom.

A pseudodifferential grounded differential amplifier (Gm_main stage of Fig. 4.11) has minimal headroom requirements. However, as shown in the plot of transconductance versus incoming voltage in Figure 4.11, such a

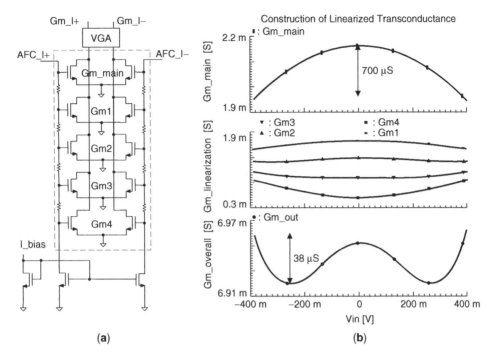

(a)   **(b)**

Figure 4.11   (a) Highly linear open-loop transconductor linearization technique utilized in LOGEN block of CMOS WLAN radio. (b) Obtained G_m for main stage, linearization stage, and overall block.

stage exhibits large variation in transconductance as a function of the incoming voltage. In this example, a variation of 700 μS is observed over an input signal voltage range of –400 to + 400 mV, with a peak transconductance of 2.2 mS. By utilizing a derivative of the technique introduced for power amplifiers in Figure 4.7, additional grounded pseudodifferential amplifier stages Gm1 through Gm4 can be generated. Each stage operates with the same incoming AC signal, but with a shifted operating point implemented through DC *IR* drops (as shown in Fig. 4.11a), additional transconductances Gm1 through Gm4 of Figure 4.11b can be generated. These additional transconductances have their peak transconductance placed at a different input differential voltage. When all of these transconductances are added at the drains of the differential pair transistors, the much more linear transconductance curve (Gm_overall) of Figure 4.11b would be generated. Now over the same input voltage range, an overall variation of 38 μS is observed with a peak transconductance of 6.97 mS. An overall transconductance variation of about 30% has been reduced to 0.5%. In this case, this reduction in transconductance variation results in more than 20 dB improvement in HD3 performance. The improvement in HD3 performance for this linearization scheme can be solved for mathematically; however, this is fairly algebraically involved and will not be presented here.

As described here, the technique is applied to a low frequency application. However, as described earlier in the case of a CMOS PA, the general technique is quite applicable to very high frequency analog and RF circuits.

CHAPTER 5

Calibration Techniques

The radios and their internal blocks must be built with adequate calibration and programming capabilities to allow them to adjust for optimum performance. Autocalibration can be used to correct and control the radio performance over temperature and process corners. Auto calibration can be used to fix the bandwidth and center frequency of the filters by adjusting the *RC* constant on the chip. Autocalibration can also help in calibrating and setting the VCO frequency in the center of its compliance range to achieve higher margins over process and temperature. In addition, autocalibration can be used to adjust overall gain in the system. Figure 5.1 is a block diagram of a robust radio with built-in autocalibration and programming capabilities (Rofougaran et al., 2005). A more specific example of a radio that can be calibrated for receive and transmit quadrature imbalances, transmit LOFT, and receive DC offsets is seen in Figure 5.2 (Bouras et al., 2003). Some of the details of this calibrated radio will be discussed shortly.

Depending on the type of autocalibration and the specific requirements of the system, calibration can be done at the factory and stored in nonvolatile memory. However, this adds to manufacturing cost and is not the preferred method. Preferably, autocalibration can be performed at chip power-up and periodically thereafter as necessary. Autocalibration may be self-contained or require the assistance of the baseband DSP.

Many RF and analog blocks would have a calibration unit built in. In addition, many radio elements can be programmed and calibrated. Many of the radio blocks may be monitored with small sensor blocks such as ADCs. Each block's performance could be measured and evaluated through an algorithm, then tuned to achieve the block's best performance. Some of the calibration algorithms may need to be fast to calibrate within microseconds on a packet-to-packet basis.

In addition to self-contained autocalibration, DSP-assisted calibration can be performed through interfacing with the baseband. Both of these methods would assist with the system's adaptiveness and robustness. Assisted calibration can initially be used to set the radio to its optimum performance by selecting the right bias current in the blocks and adjusting the on-chip frequency tuning. DSP-assisted calibration can also improve the image

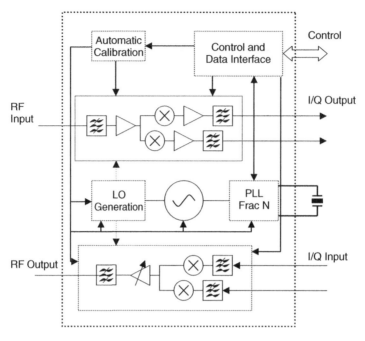

Figure 5.1 Block diagram of general radio with built-in autocalibration and programmability to ensure high performance.

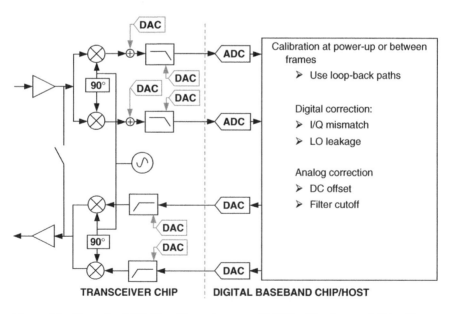

Figure 5.2 Example of 802.11a with quadrature and LOFT calibration capabilities (Bouras et al., 2003).

rejection on chip by measuring the *I/Q* channel imbalance and adjusting the phase and amplitude on the *I/Q* signals. Another important use of programmability is its ability to adjust the radio blocks to their optimum performance based on feedback from the processed radio data with a special algorithm from the baseband. This adjustment can be very critical in obtaining a functioning radio for harsh environments where interference, jamming, and other noise can degrade radio performance. Although calibration will for the most part be completed at the start of the communication, a few calibrations also take place during the on-going communication and can improve system performance by the radio "adapting" to its environment.

Autocalibration, whether DSP assisted or self-contained in the radio, would allow the inherent tolerances required of the radio to be significantly relaxed. This would result in much higher yields, which in turn would result in lower cost. At the same time, autocalibration would allow for better performance than available from the uncalibrated radio. The utilization of such calibration algorithms is increasingly more important in future generation radios.

Several calibration techniques will now be examined. The details of each calibration technique can occupy a whole book chapter. However, we will discuss some very briefly and others in some detail.

5.1 VCO CALIBRATION

As described earlier a low VCO gain (K_{vco}) is desirable in order to maintain a good PN and high rejection of noise coupling on the VCO control line and the supplies. In order to accommodate a low K_{vco} and at the same time be able to cover a wide bandwidth, banks of switch capacitors need to be utilized. In order for a proper operation, for a desired frequency band, the proper switched capacitors are enabled such that the VCO control line is set to about the midpoint of the charge pump compliance range (typically about supply voltage $V_{dd}/2$) for the PLL to lock. The switch capacitor banks can be linearly weighted or binary weighted. An example of VCO tuning curves as a function of the VCO control voltage with the various switch capacitor codes is given in Figure 5.3. This VCO tuning system is designed for a radio that covers the lower U.S. 802.11a band. Clearly, in order to enable locking across the entire worldwide 802.11a bands while maintaining the same VCO gain, many more switched-capacitor banks would be required.

Due to process, temperature, and supply variations, it is not sufficient to simply set the required switched-capacitor banks for a given channel based on a simple look-up table. It is essential to use an autocalibration loop to enable proper operation over these parameters. Various algorithms can be uti-

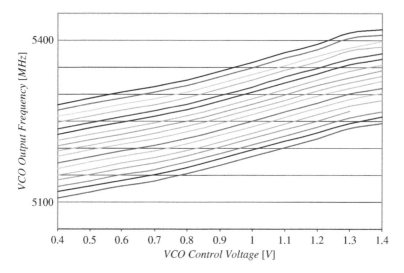

Figure 5.3 VCO tuning curves covering lower U.S. 802.11a band. The calibration algorithm selects the optimal curve such that the VCO can operate at a voltage around the middle of the charge pump compliance range.

lized for autocalibrating the VCO switch capacitors. These include open-loop calibration, closed-loop calibration, or a combination of the two. Each scheme has its advantages and disadvantages (e.g., the calibration settling time would vary depending on the type of calibration).

In a closed-loop calibration scheme, the PLL is programmed to the desired channel and the PLL loop is engaged in the normal fashion. Then a binary or linear search algorithm is engaged which enables the various switch-capacitor networks. After each setting, the control voltage of the VCO is monitored. If the control voltage falls between a predetermined set of values as determined by a pair of comparators, then the desired calibration value is achieved. Otherwise, the next search step is enabled and the procedure is repeated until the algorithm is successful in finding the proper code for the switch-capacitor network. A closed-loop calibration scheme is usually quite accurate but would take a long time to converge since for every step the PLL is required to settle.

Alternatively, an open loop calibration scheme can be used. In this scheme, the PLL is disengaged, the PLL dividers are properly programmed for the desired frequency channel, and the control voltage is forced to a desired voltage (usually the midrange of the charge pump compliance voltage). Then, for predetermined reference divider cycles, the number of the VCO divided output transitions is counted. The switched-capacitor code is changed via a linear or binary search algorithm until the desired count value is achieved. The advantage of an open-loop calibration algorithm is that it is quite fast since it does not have to wait for the PLL to settle after each code change. However, it may not be as accurate as the closed-loop calibration.

Clearly, a combination open–closed loop calibration is possible that would provide a compromise between calibration speed and accuracy.

5.2 AUTOMATIC FREQUENCY CONTROL

A receiver that has its final stage centered at DC would have to be able to reject DC offsets which would result from self-mixing of the receive mixer or receiver baseband mismatches. In order to accomplish this some form of a high pass filter is often used.[76] For example (Behzad et al., 2003), there are three high pass filters distributed in the receive path. The poles of these high pass filters cannot be too low in frequency as they would result in long transient settling during gain changes or TX-to-RX switching. On the other hand, the poles cannot be placed too high in frequency as they will attenuate the lowest OFDM subcarriers and cause significant group delay variation and therefore cause a degradation in the system performance. In the ideal case and during steady-state operation, the poles in the receiver described in this example are placed at about 120 kHz. Figure 5.4a. shows an OFDM-modulated signal with its 52 subcarriers and their payloads. The lowest subcarriers are at ±312 kHz, and the highest subcarriers are at ±8.125 MHz. In the ideal case, as shown in Figure 5.4a, none of the subcarriers are attenuated by the filtering in the receive chain. However, since the standard requires only a 20-PPM crystal, a total of 214 kHz of frequency offset (107 kHz RX plus 107 kHz TX) can be present in the received spectrum relative to the location of the baseband filter poles. This would result in the lowest frequency subcarrier as well as possibly the highest subcarriers to get heavily attenuat-

---

[76]Other alternatives also exist. For example, DC offset cancellation DACs can be employed at various points along the receive chain. There are advantages and disadvantages to this technique. One advantage is that an AFC circuit would not be necessary. Further, the inner tones of the OFDM-modulated signal would not take a hit due to amplitude roll-off or group delay variation. Another advantage of this scheme is that the receiver is capable of settling quickly in response to any transient changes (e.g., gain change). On the other hand, after an initial calibration and possibly periodic calibrations thereafter, the cancellation is performed open loop and would need to track the gain settings accurately. Further, the offset cancellation DACs will need to be designed to have low noise levels in order to avoid an SNR hit due to these blocks. Finally, for a multistage DAC-based DC offset cancellation scheme requires carefully developed algorithms to ensure that the proper offset levels are applied at the proper points.

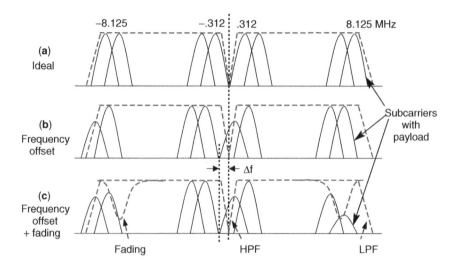

Figure 5.4 (a) Baseband spectrum of received and properly centered 802.11a signal. HPF and LPF corners are shown. (b) Same spectrum after being subjected to Δf frequency offset. (c) Spectrum of (b) additionally subjected to multipath fading.

ed by the receive filters (Fig. 5.4b). This can have a severe impact on the EVM and PER of the received signal. The situation can be even worse in the presence of a multipath channel (Fig. 5.4c). Under a multipath environment, many of the subcarriers are heavily attenuated by the channel, and as a result the receiver has to rely on the none-attenuated subcarriers more heavily. When these subcarriers are subject to the heavy receive filtering, overall system performance could degrade quite severely.

Most receivers require an automatic frequency correction in order to correct for frequency offsets. However, the correction is typically done entirely in the digital domain and will therefore not be able to avoid the filtering problem described above. In the implementation described here, the system corrects for the frequency offsets of the receiver and transmitter in the analog domain and therefore eliminates the filtering of the desired OFDM subcarriers. The estimation of the frequency offset and the generation of the correction baseband frequency, however, are done in the digital domain. Therefore this operation takes place in both the analog and the digital domains [hence the term *mixed-mode automatic frequency control* (AFC)].

The block diagram of the mixed-mode AFC system is shown in Figure 5.5. During the receive cycle, the received in-phase and quadrature-phase signals are applied to the baseband ADC. A frequency estimation of the effective frequency offset (Δf) is made (this estimation includes frequency offsets from the transmitting source as well as that from the receiver itself).

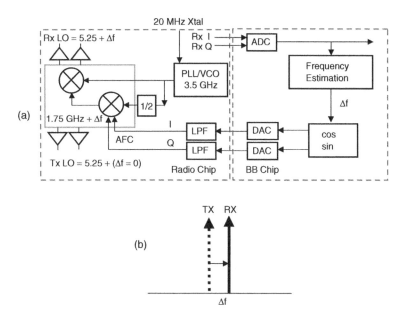

Figure 5.5 Block diagram of mixed-mode automatic frequency correction (AFC) loop utilized in 802.11a system. The correction frequency Δf is only applied during RX mode. Therefore the effective LO frequency is "hopping" (Behzad et al., 2003).

As a result, correction frequency signals $\sin(\Delta f)$ and $\cos(\Delta f)$ are generated in the digital domain and passed on to high precision, low offset DACs. These signals are then passed on to the radio chip where they are filtered. The AFC I and Q signals are then applied to the first set of mixers in the LO generation block. Assuming a VCO frequency of 3.5 GHz, the first mixer would have an output image-rejected frequency of 1.75 GHz + Δf. This signal is then mixed with the VCO frequency in the second set of LO generation mixers, resulting in the output frequency of 5.25 GHz + Δf. In the transmit mode, Δf is set to zero, and an output LO of 5.25 GHz would be generated. The LO generation signal, therefore, is constantly switching back and forth between the TX and the RX frequencies as shown in Figure 5.5b. Also note than in order to reject the image of the AFC tone, the entire AFC signal path is implemented in the complex form (I and Q). In summary, the relations for the RX and TX LO outputs are

$$f_{\text{LO\_TX}} = \tfrac{3}{2} f_{\text{vco}} \qquad f_{\text{LO\_RX}} = \tfrac{3}{2} f_{\text{vco}} + \Delta f_{\text{afc}}$$

The correction frequency Δf is no larger than 240 kHz.

It is important to note that in a 802.11a system the AFC correction must occur on a per-packet basis and therefore needs to occur during the preamble processing and needs to be quite fast. This fact along with potential spur problems rules out a fractional synthesizer as an alternative solution. Furthermore, unlike a cellular system in which the transmitter (a base station) has excellent frequency accuracy, an 802.11 access point is subject to the same frequency tolerance (20 PPM) as a client. Therefore, the receiver has to correct for the frequency offsets of its own receiver *and* the transmitter. Since the transmitter can change from packet to packet, a static frequency correction similar to the one used in cell phones (using a VCXO) is not a feasible solution for the 802.11 system.

The simplified block diagram of the LO generation mixer is shown in Figure 5.6. The filtered AFC signals are applied to the transconductance stages (Gm1 to Gm4) of the mixers. The resulting signals are gain controlled through current-mode gain control blocks VGA1 to VGA4. The gain control is placed *after* the transconductance stage so that the DC offsets of the transconductance stage scale with gain control. The resulting signals are unconverted with an LO of $0.5f_{vco}$ and then further unconverted with an LO of f_{vco}, resulting in the desired $1.5f_{vco} + \Delta f$ output frequency.

Ideally, at the output of the first set of mixers, a single tone at $0.5f_{vco} + \Delta f$ would be present. However, in reality, due to the nonideal image rejection of the mixers, an image tone would also be present at $0.5f_{vco} - \Delta f$. Additionally, due to the DC offsets at baseband, an LOFT term exactly at $0.5f_{vco}$ would be present. Finally, due to the nonlinearities of the transconductance

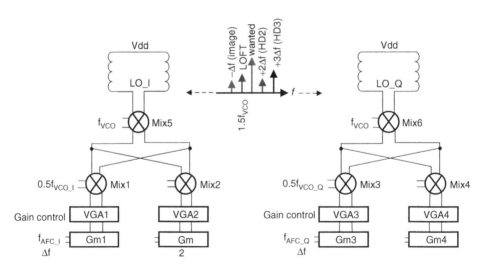

Figure 5.6 Simplified block diagram of LO generation mixer utilized in mixed-mode AFC loop.

stage, third- and possibly second-order harmonics of the AFC signal will be present at $0.5\,f_{vco} + 3\,\Delta f$ and $0.5\,f_{vco} + 2\,\Delta f$. All these spurs will be unconverted to the final 5 GHz frequency by the second set of mixers (mix 5 and mix 6). The challenge in the design of the LO generation mixer is to generate output signals large enough for the receive and transmit path mixers over temperature and process variations, while maintaining these spurs to less than 40 dB below the desired signal. The simplified schematic of the inphase side of the LO generation "stacked" mixers is shown and discussed in Figure 4.10. A similar block generates the desired quadrature-phase output.

An AFC self-calibration mode is integrated in the system by which the LOFT of the AFC block as well as the image term in the AFC block can be automatically canceled at startup. In this calibration mode, a test AFC signal is generated and applied to the receiver RF as well as the LO paths. The down-converted signal is sampled by the ADCs and analyzed. The correction quadrature terms and DC terms are then applied to the AFC DACs in order to obtain a pure AFC-shifted LO signal at the receiver mixer port.

It is important to note that a mixed-mode correction is most important in the presence of large crystal frequency offsets. If high quality (low PPM variation) crystals can be guaranteed in the system, the AFC block can be eliminated with minimal performance penalty. Low PPM crystals are typically available at some additional cost. So the decision to utilize mixed-mode AFC circuitry is a trade-off between bill-of-material cost, design complexity, and desired performance levels. Note that the quality of the crystal on both the receive side as well as the transmit side must be considered in making this decision.

The performance of the receiver in the presence of frequency offsets and multipath distortion is examined in Figure 5.7. As shown in Figure 5.7a, due to the effect of the severe multipath channel of 200 ns RMS delay, –2 dBc multipath signal strength, and 180° of relative phase, the received spectrum displays deep nulls. Also, the multipath channel results in an increase in the error vector magnitude of the OFDM subcarriers that falls at the frequency of the nulls. However, the more severe effect on the error vector spectrum is due to the large 200-kHz frequency offset that is present on the received baseband signal (relative to the location of the receive high pass filter poles) which causes severe attenuation of the lowest frequency subcarriers. This results in a very large increase in the EVM of the lowest frequency subcarrier (subcarrier –1). On average the EVM has increased from less than 5% to over 24% for this subcarrier, and the peak EVM has increased to over 60% for this subcarrier. This results in the outlier points in the constellation diagram of Figure 5.8a. The average EVM in this case is –24.3 dBm, which is not sufficient for good PER performance.

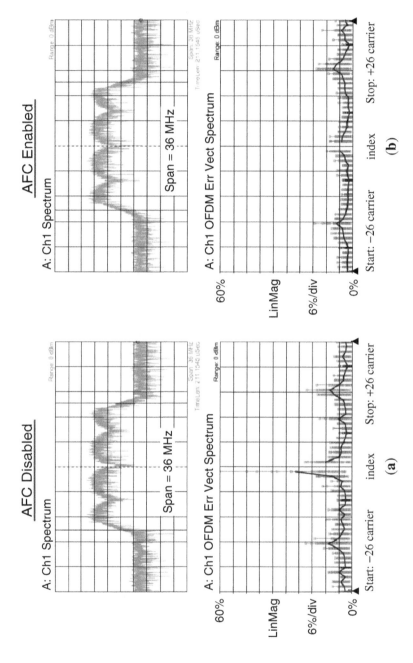

Figure 5.7 Spectrum and EVM plots of 802.11a receiver in presence of multipath channel (200 ns rms delay, −2 dBc multipath signal strength, 180° phase) as well as total LO frequency offsets of 240 kHz with AFC loop (**a**) disabled and (**b**) enabled.

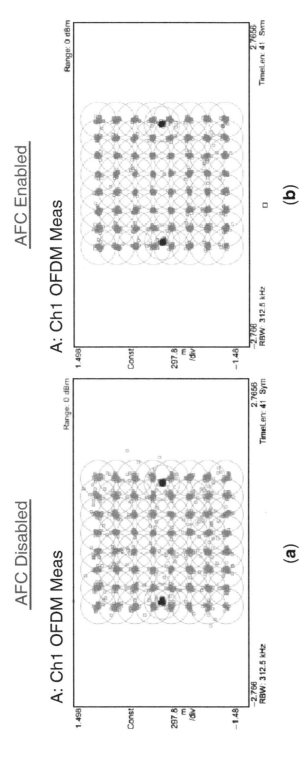

Figure 5.8 Constellation plots of same receiver (**a**) with and (**b**) without AFC enabled. With AFC enabled, the EVM has improved by more than 4 dB, enabling operation and low PER at 54 Mbps.

171

Now with the AFC enabled, as shown in Figure 5.7b, the impairments due to the multipath channel are still present, but the effects of the frequency offset have been significantly reduced as apparent in the error vector spectrum plot of Figure 5.7b. In this case most of the constellation diagram outlier points have been eliminated, and even in the presence of a very severe multipath channel, a receive EVM of −28.4 dB is achieved (Fig. 5.8b). This EVM is more than sufficient for reliable 54-Mbps performance. It is also important to note that while in the presence of relatively mild multipath channels, the AFC would improve high data rate sensitivity level, but in the case of a severe multipath channel (such as the example described here) and without AFC, the receiver PER floor may be raised to a level at which high data rate operation may not be possible at all.

5.3   QUADRATURE ERROR AND LOCAL OSCILLATOR FEEDTHROUGH CALIBRATION

As discussed earlier, one of the main analog impairments impacting the performance of high order modulation systems is quadrature imbalances on the transmitter and the receiver. Traditionally, a designer would utilize good design and layout matching techniques to achieve good IQ imbalance. More recently, some newer architectures have been proposed that inherently possess a higher tolerance to quadrature imbalances. At times, the systematic mismatches are characterized in the lab, and a fixed correction is applied in the digital domain in the form of pre- and/or postdistortion coefficients (using a scheme known as a Gramm–Schmidt orthogonalization method). The problem with this approach is that it cannot account for process, voltage, and temperature variations. Despite all these techniques, IQ imbalance remained as one of the main system performance bottlenecks and production yield issues. The most recent developments, however, enable the calibration of IQ imbalances. Many methods have been proposed. Some utilize an entirely analog approach. The most powerful and common techniques utilized to combat IQ imbalances in WLAN systems, however, operate on a mixed-mode (analog-and-digital) scheme.

One such mixed-mode scheme is utilized in the transceiver of Figure 5.2 (Bouras et al., 2003) based on an algorithm proposed by Cavers (1997). The transmit quadrature error calibration mode is shown. An envelope detector is utilized to measure the quadrature imbalance at some point after the quadrature up-conversion stage. Ideally, the envelope detector would generate a tone at twice the frequency of the applied baseband tone at the output of the envelope detector if an image is present (i.e., quadrature imbalance exists). This information can then be utilized to "predistort" the digital baseband signal to correct the *IQ* imbalance in the system (typically introduced by the RF LO generation circuitry). In this implementation, the output of the envelope detector is fed to the input of one of the receive signal path ADCs and passed on to the digital PHY section for predistorting the transmit digital data. These calibrations can be performed at system startup and periodically thereafter as required.

Calibrating of the *IQ* imbalance in the digital domain by utilizing predistortion is a very powerful technique for calibrating any reasonable amount of transmit quadrature imbalance (although certain limitations should be considered in the design of the calibration circuitry and the dynamic range of the ADCs and DACs).

Once the transmitter quadrature errors are corrected for, the signal is looped back as shown in Figure 5.9. The looped-back signal is down converted through the receiver quadrature down converters, passed through the baseband receive analog circuitry, and then sampled by the ADCs. The signal is then analyzed in the digital domain, and the proper postdistortion *IQ* calibration coefficients are calculated and applied to the received signals during normal operation. Figure 5.10 displays the constellation diagram of a received (non-OFDM) QPSK signal before and after the quadrature correction is applied to the receiver. Note that no frequency locking has been applied to the demodulated signal (hence the "circular" constellation pattern). For a perfect quadrature, a perfect circle centered at coordinates (0,0) should be observed. As can be seen in this figure, in the presence of quadrature imbalances, the constellation diagram has an oval shape and is not centered at (0,0).

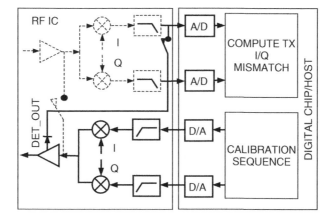

Figure 5.9   Radio of Figure 5.2 shown in TX calibration mode (Bouras et al., 2003).

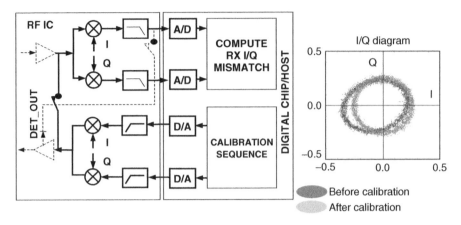

Figure 5.10   Radio of Figure 5.2 shown in RX calibration mode. The received constellation diagrams for a received QPSK signal before and after calibration are also shown (Bouras et al., 2003).

It is important to note that most often the *IQ* imbalances in a radio are caused by the LO generation circuitry. As such, the *IQ* imbalances are often constant across the bandwidth of the received signal and a fixed pre- or postdistortion phase and amplitude correction terms are sufficient to calibrate the system. If the *IQ* mismatches, however, are caused by the baseband circuitry, a (baseband) frequency-dependent *IQ* calibration is likely to be required. The estimation methodology outlined above, however, is still applicable. Multiple test tones at different baseband frequencies, however, would be required.

The same envelope detector used for estimating the *IQ* imbalances on the transmitter can be utilized to measure the magnitude of the LOFT as well. Ideally, the envelope detector output would show a tone at the frequency of the applied baseband tone if LOFT is present in the transmitter output. Once the magnitude of the LOFT is estimated, a calibration algorithm can cancel the LOFT by applying the proper DC voltages or currents in the transmitter baseband (digital or analog path). Implementing the LOFT compensation entirely in the digital domain may seem attractive. However, such a scheme can pose certain system level problems, especially if a significant amount of gain control is required at the transmitter baseband.

Other schemes can also be utilized for the detection of the LOFT. For example, the power detector which is typically integrated in the transmit chain to monitor the transmitted output power of the transmitter can be used to estimate the power of the LOFT tone when no other signals are applied at the transmitter input. A search algorithm can then be applied with the goal of reducing the power detected by the power detector. This scheme can be applied to applications where a relatively loose LOFT rejection is acceptable in the system. The degree of LOFT rejection in this approach is typically limited by the dynamic range of the power detector.

Because of the importance of this topic, we will consider a more detailed implementation of a high performance transmitter with quadrature and LOFT cancellation schemes as our second case study in Chapter 6. The implementation presented in Chapter 6 has the additional advantage that the LOFT cancellation can be maintained to a high degree even as the baseband transmitter gain control is varied.

5.4 BIAS CURRENT CALIBRATIONS (*R* CALIBRATION)

Using certain reference voltage generation circuitry, such as bandgaps, relatively constant voltages on the chip can be synthesized. These voltages have little variation as a function of process, temperature, and supply voltage. Typically, however, in order to generate a current, these voltages are presented across a resistor to generate a bias current. Any current generated using this scheme would inherently possess the process and temperature variations of this resistor. Modern CMOS process technologies offer unsilicided polysilicon resistors with reasonably low temperature coefficients (often P-type unsilicided polysilicon with a temperature coefficient of approximately 400 PPM). However, the foundries typically specify a process variation of ±30%.[77] This would indicate that even if the reference voltage is perfect and has no variations over process, the generated bias current would have a variation of ±30%. This is a very large variation and unacceptable for many applications.

One solution would be to reference the on-chip generated voltage on an a tight-tolerance external (on-board) resistor to generate the desired bias current. This scheme has the limitation that every such current would require a separate pin and an external resistor. Alternatively, a calibration scheme can be used where the generated current based on an on-chip resistor is compared against the current generated with the same reference voltage but with an external tight-tolerance resistor. The calibration algorithm can then attempt to equalize the currents by switching in the proper fractional currents or adjusting the on-chip resistors until the currents are equal. The obtained calibration code can then be applied to all the current generation circuitry on chip. Utilizing this scheme, bias currents with variations in the few percent range can easily be generated.

---

[77]Often the real 3-sigma variation is significantly smaller than this number, but the foundries would not guarantee this!

This calibration technique works quite well for circuits in which the generated current is (inversely) linearly related to the value of the resistor such as a bandgap and proportional to absolute temperature (PTAT) currents. At times, bias currents may be needed in which the generated current is not linearly related to the value of the resistor. One common example is a MOS G_m-based bias circuit, in which case the bias current is inversely related to the MOSFET G_m. In this case the generated bias current is related to $1/R^2$. A small variation in the value of R therefore can result in a large variation in the bias current. The application of a calibration circuit to this case is even more crucial that the previously discussed cases. However, it should be noted that a small residual calibration error can result in fairly large bias current error.

One important point is that if a constant voltage swing is desired from a certain resistively loaded stage (e.g., on a high frequency divider), a calibrated current must *not* be used, assuming that the stage is utilizing the same type of resistor as that used for generating the bias current. In such a block the variation of the resistor is canceled in the generation of the output voltage.

5.5   FILTER TIME-CONSTANT CALIBRATION (*RC* CALIBRATION)

Many stages of filtering are often used in transceivers. For example, a baseband receive filter is typically used in WLAN systems to reject any adjacent channel interferers. In this case, if the filters are too narrow, the desired channel would get prematurely cut off. This in turn would result in degraded system performance due to the attenuation of the higher order OFDM subcarriers as well as group delay variation introduced close to the filter corner. On the other hand, if the filters are too wide, sufficient rejection of the adjacent channel interferer may not be achieved. It is therefore important to be able to achieve the proper filter bandwidths despite process and temperature variations. The source of filter bandwidth variations in a filter is the variation of the resistors and capacitors used in the filter.[78]

Various filter time-constant calibration techniques have been proposed and used. Most of these techniques rely on off-line calibration where the filter is calibrated at startup or periodic intervals when it is not in use. Further,

---

[78]Resistors and capacitors are used to establish the poles and zeros in an "opamp-*RC*"-based filter. Alternatively, transconductances and capacitors are used to establish these poles and filters in "gm-*C*" type filters. Clearly the master blocks used for calibrating these blocks need to be of the same kind of basic building blocks in order to achieve the desired results. We specifically discuss an *RC*-based calibration approach. A similar gm-*C* based approach can be used where appropriate.

most of these techniques rely on a replica (master) *RC* block that is calibrated. The filter (slave) block is built with the same type of resistors and capacitors as the master block. Once the master block is calibrated and the tuning code is obtained, the same tuning code is applied to the slave (filter) block.

One such approach is to generate a master *RC*-based relaxation oscillator. The number of cycles "counted" by this *RC* oscillator is then compared to a certain number of clock cycles of a known high accuracy clock (typically the crystal oscillator circuit which is available on most radios). Switched capacitors or switched resistors are then turned on or off in order to calibrate the *RC* oscillator cycle counts against that of the crystal oscillator. The calibration code is then achieved and applied to the slave (filter) block(s).

5.6 OTHER CALIBRATIONS

Various other calibration schemes can be utilized in a radio as necessary. For example, many transmitters integrate a power detector at the final gain stage output. This enables the system to be able to calibrate out gain variations over process, voltage, and temperature to maintain the desired output power level. In particular, this is important for CMOS-based transmitters which can have a significant amount of gain variation over temperature. This is because not only the Q of the inductors is significantly degraded over temperature, but also the transconductance of the CMOS devices takes significant hit at hot temperatures.[79] Of course, the accuracy of the power level calibration is dependent on the accuracy of the integrated power detector. Integrated power detectors have their own sources of inaccuracies. Certain calibration techniques can be applied to correct for the inaccuracies of the power detectors.

Another example of a block that can be utilized for automatic calibrations on the radio is an integrated temperature sensor. Typically a relatively coarse temperature measurement is sufficient to correct for many issues that may arise over temperature. If a finer resolution temperature measurement is required, a high accuracy temperature sensor would be required. The design of high accuracy temperature sensors can be challenging. A temperature sensor can be used, for example, to compensate for the variation of the IQ compensation over temperature. As a further example, a temperature sensor can also be used to determine when a VCO recalibration may be required.

---

[79]For bipolar devices, the transconductance is inversely proportional to the absolute temperature. Therefore, by using a PTAT-based biasing, the transconductance of the bipolar device can be kept constant over temperature.

CHAPTER 6

Case Studies

Now that we have discussed the 802.11 system requirements, understood analog impairments that can impact the system performance, and looked at various calibration techniques, it is time to look at two case studies. In the first case study we will discuss a 802.11a radio implementation in some detail (Behzad et al., 2003). In the second case study we will look at a high-performance WLAN transmitter utilizing quadrature and LOFT calibration algorithms (Lee et al., 2006). These case studies allow us to tie in the various topics that have been discussed together and represent how these issues are taken into account in the design of a transceiver.

6.1 CASE STUDY 1: A CMOS 802.11A TRANSCEIVER

6.1.1 Architecture and Circuit Implementation

At the time of design, the goal of this work was to achieve the lowest cost, highest performance, and lowest power consumption radio for the 802.11a standard. It was therefore decided to use a direct-conversion architecture without requiring any external filters. To allow for low cost and future integration with DSP, a CMOS process was chosen. Furthermore, an integrated PA was added to the transmitter to reduce the bill-of-materials cost. Finally, extensive use of autocalibration schemes enhanced performance and increased yield.

However, these choices result in a host of challenges that need to be dealt with in the architectural implementation and/or the circuit design of the blocks. For example, the choice of a direct-conversion architecture with an integrated power amplifier requires the designer to deal with the following issues:

- DC offsets which result from self-mixing of the receive mixer as well as DC offsets which result from baseband block mismatches and the high gain of the baseband stages. If these DC offsets are not properly rejected, they will result in clipping of the subsequent stages.

Wireless LAN Radios: System Definition to Transistor Design. By Arya Behzad
Copyright © 2008 the Institute of Electrical and Electronics Engineers, Inc.

- Flicker noise on the receive path can impair the SNR of the lowest index OFDM subcarriers.
- The receive baseband path can have potential oscillation problems due to the fact that most of the receive path gain is implemented at a single frequency (baseband).
- The transmit path can have potential oscillation problems since large amounts of gain are required at the 5-GHz RF.
- LO pulling issues by the on-chip PA can cause system problems.
- LOFT issues on the transmitter have to be dealt with.
- The on-chip power amplifier is required to have a very high linearity in order to be able to accommodate the large PARs present in the OFDM signal.

Among the autocalibration schemes that are incorporated in this radio, the AFC is the most challenging to implement (the AFC was discussed in some detail in Chapter 5).

The simplified radio architecture is shown in Figure 6.1. The transceiver consists of a receiver (RX), a transmitter (TX), a frequency synthesizer, a high-speed custom JTAG digital control block, and calibration blocks for bias currents (I_{bias}) and RC time constants. A fully differential signal path is

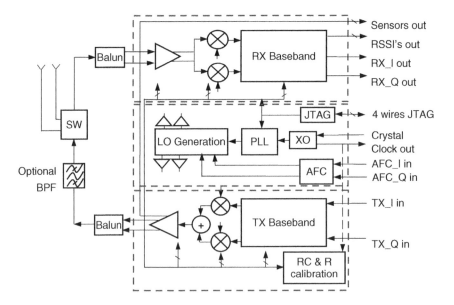

Figure 6.1   Simplified high level block diagram of transceiver of Case Study 1 (Behzad et al., 2003).

used throughout the transceiver to reduce LOFT, LO leakage, and common-mode noise. The I_{bias} calibration block uses an external 1% resistor to generate a code for various bandgap circuits, resulting in biasing independent of on-chip resistance variations. The RC time constant calibration circuit uses an RC-based oscillator with a reference provided by the crystal oscillator to generate a 5-bit time constant code for the on-chip filters. RC calibration is performed at startup only since the variation of the resistors and capacitors utilized here with temperature is small. An integrated temperature sensor allows the DSP to make optimal system level adjustments as a function of temperature. The JTAG block controls all of the radio functions, including receive and transmit gain controls.

The transceiver, as shown, along with a companion baseband PHY/MAC chip, two external baluns, an optional bandpass filter (BPF), and a transmit-receive (TR) and/or diversity switch constitutes a full 802.11a system. We will now discuss each building block in more detail.

6.1.2   Receiver

The receiver is shown in detail in Figure 6.2. The RF signal is amplified first by an on-chip tuned LNA and then down converted to baseband by two quadrature mixers. The output of the mixers is fed into the first high pass variable-gain amplifier (HPVGA). The output of the first HPVGA is fed into a fifth-order Chebyshev low pass filter (LPF) for channel selection. The output of the LPF is then fed to second and third HPVGAs. The outputs of the last HPVGA are passed on to ADCs on the baseband chip. The entire signal path is differential but is shown as single ended after the LNA for the sake of simplicity. The receiver front end includes on-chip matching for the 5-GHz input. A high gain, low noise, high linearity, and gain controllable

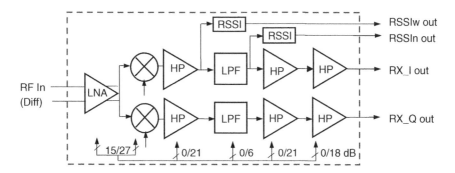

Figure 6.2   Simplified block diagram of receiver of Case Study 1. All signal path circuits are implemented in a fully differential topology (single ended shown to simplify diagram).

front end allows for optimal system trade-offs between sensitivity and linearity. The RX gain is carefully distributed before and after mixers to minimize the $1/f$ noise contribution to the overall system NF. Three stages of HPVGAs are incorporated throughout the baseband signal path and provide both high gain and DC offset rejection. These HPVGAs are programmable in 3-dB-gain steps from 0 to 21 dB (0 to 18 dB for the last HPVGA). Since the preamble duration is 16 μs, the DC offset cancellation has to be very fast, and at the same time, it must not attenuate the lowest subcarriers of the signal. Therefore the HPVGAs are designed to accommodate dynamic DC offsets which result from gain changes. A fifth-order Chebyshev LPF is integrated between the first and second HPVGAs that acts to reject any CW or modulated interferers. The LPF is automatically calibrated within ±2% tolerance to ensure precise channel selection. Dual receive signal strength indicators (RSSIs) are integrated before and after the LPF. These RSSIs allow for the system to determine whether a received signal strength is dominated by an out-of-band interference signal or by an in-band desired signal. For example, if the second RSSI output voltage is smaller than the first, this will be an indication that the large incoming signal is due to an out-of-band interference. Based on this determination, optimal front-end gain can be set for the proper trade-off between sensitivity and linearity.

The receiver NF and gain (as measured with a noise head) as a function of the baseband frequency are shown in Figure 6.3a. Note that the x axis is in the logarithmic scale. An average NF of better than 4 dB in the receive

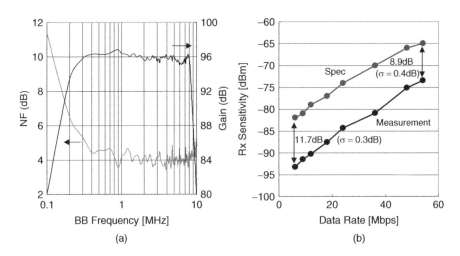

Figure 6.3   Measured receiver gain and NF versus frequency (**a**) and sensitivity versus data rate (**b**).

path is achieved. It is also important to note that the NF at the frequency of the lowest frequency subcarriers are maintained to better than 5.5 dB, allowing for these subcarriers to maintain high SNR. The overall receiver transfer function with both high pass and low pass characteristics is apparent in the gain plot of Figure 6.3a. The receiver sensitivity is shown in Figure 6.3b along with the 802.11a requirement as a reference. For the lowest data rate of 6 Mbps, a typical sensitivity of –93.7 dBm and a standard deviation of 0.3 dB are achieved. At the highest data rate of 54 MBps, a typical sensitivity of –73.9 dBm and a standard deviation of 0.4 dB are achieved. All sensitivity measurements are obtained with the actual baseband companion chip to this RF transceiver operating with a hard Viterbi decoder. A soft Viterbi decoder in the digital baseband can offer several decibels of improvement in the sensitivity of the system at the higher data rates.

6.1.3 Transmitter

The transmitter block diagram is shown in Figure 6.4. Like the receiver, the transmitter is based on a direct-conversion architecture and is fully integrated on-chip. It incorporates third-order Butterworth LPFs which receive the signals from the baseband I and Q DACs. The outputs of the LPFs are then applied to baseband VGAs. The signals are directly up converted to the RF and combined before a RF VGA. The signal is then amplified by a power amplifier driver (PAD) and applied to a high linearity, high power integrated class AB power amplifier. On-chip matching is provided for the PA. The transmitter incorporates baseband and RF gain control for optimum trade-off between linearity, noise, LOFT, IQ balance, and power consumption. A gain control of 33 dB is provided at the baseband, 3 dB at the mixer, and 35 dB at the RF amplifiers. Some gain control is needed to compensate for power variations resulting from changes in process, temperature, and power supply. The gain control is also used to allow for transmit power control for

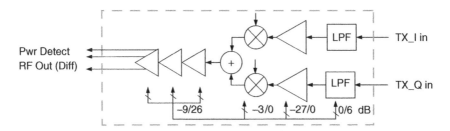

Figure 6.4 Simplified block diagram of transmitter of Case Study 1. All signal path circuits are implemented in a fully differential topology (single ended shown to simplify diagram).

the future versions of the 802.11a standard. An integrated power detector allows for the DSP to set a constant output level in the presence of process, temperature, or supply variations. A scheme for the cancellation of the LOFT is also integrated in the transmitter. This scheme utilizes DACs at the mixer transconductance stage to cancel out the LOFT at RF. A transconductance linearization technique similar to that discussed in Chapter 4 is utilized in the PA driver and the PA of this transmitter. The transmitter shown in Figure 6.2 achieves a 1-dB compression point of +19 dBm and a saturated output power of +23 dBm.

The measured transmit output power versus the data rate is shown in Figure 6.5. At the highest data rates associated with QAM-64 modulation, the maximum transmit power is limited to 12.8 dBm by the required EVM and the high PAR of the OFDM signal. Under these conditions, the integrated PA consumes 63 mA from 3.3 V and has an efficiency of 9.2%. As the data rate is reduced and the modulation is switched to QAM-16, the maximum transmit output power is limited equally by both the EVM requirements and the spectral mask requirements of the standard. For example, at 36 Mbps, a maximum transmit power of 18.7 dBm is achieved. In this case, the integrated PA consumes 77 mA from 3.3 V and has an efficiency of 32.9%. As the data rate is further reduced, the maximum transmit power can no longer be increased and is still limited by the spectral mask requirements. It is interesting to note that if the spectral mask requirements were to be ignored, at the lowest data rates, a maximum transmit power equal to the saturated output power of the

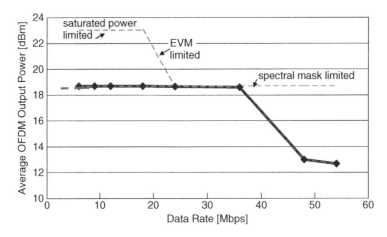

Figure 6.5 Average OFDM TX output power as function of data rate. Note that at the lower rates the maximum TX power is limited by the required EVM, whereas at the higher rates the TX power is limited by the spectral mask requirements of the standard.

transmitter (+23 dBm) could be achieved while still satisfying the EVM requirements of the standard. It is also important to note that what is shown on the y axis is the average transmit power and, that during statistical peak excursions of the OFDM signal, output power levels significantly higher than the average level (at times equal to the saturated power) are observed.

Figure 6.6a displays the measured transmitter output spectrum while transmitting a +12.8-dBm, 54-Mbps, QAM-64 modulated signal. In this case the maximum transmit power is limited by EVM requirements. In applications in which a higher 54-Mbps transmit power is required, an external PA can be used and the internal gain of this chip can be backed off to account for the gain of the external PA and the desired output power. Figure 6.6b, on the other hand, displays a +18.7-dBm, 36-Mbps, QAM-16 signal. In this case the maximum transmit power is equally limited by the EVM requirements as well as the spectral mask requirements.

The measured constellation diagram of a QAM-64, 54-Mbps, +6-dBm transmitted signal with an EVM of –33 dB is shown in Figure 6.7. Recall that the 802.11a standard requires an EVM of –25 dB, and therefore 8 dB of margin is present here. That the constellation points fall within a small radius around the center of the crossing points of the constellation circles is an indication of the high quality of the transmitted signal. This is an indication that all transmit signal impairments, including phase noise, thermal noise, nonlinearity, phase and amplitude imbalance, and group delay, are maintained to very tight tolerances. Note that the two darker constellation points on the real axis are from the pilot tones, which are always transmitted with BPSK modulation.

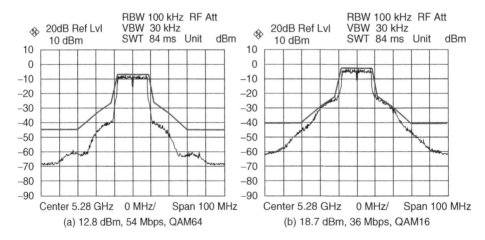

Figure 6.6 Measured output power spectrum: (**a**) EVM limited; (**b**) EVM and spectral mask limited.

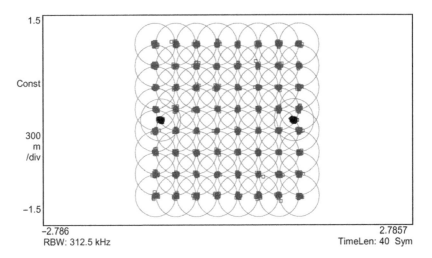

Figure 6.7 Measured constellation diagram at TX output (EVM = −33 dB, QAM64, 54 Mbps, +6 dBm).

6.1.4 Phase-Locked Loop

An external 20-MHz crystal is used with an on-chip oscillator as a reference for the integer-N PLL with programmable loop bandwidth (Figure 6.8). Various PLL-related issues typically impair the operation of a direct-conversion transceiver. These impairments include pulling effects by the on-chip PA, large in-band LO leakage, and large DC offsets as a result of LO self-mixing. In order to reduce these impairments, a "fractional VCO" is used in which the desired RF is 1.5 times higher than the VCO frequency (Darabi et al., 2001). Furthermore, in order to reduce negative effects due to the crystal tolerance, a mixed-mode AFC design is used.

The measured phase noise of the system is shown in Figure 6.9. An in-band phase noise of −100 dBc/Hz at an offset frequency of 30 kHz and with

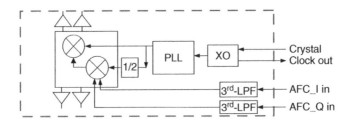

Figure 6.8 Simplified block diagram of PLL of Case Study 1. All signal path circuits are implemented in a fully differential topology (single ended shown to simplify diagram).

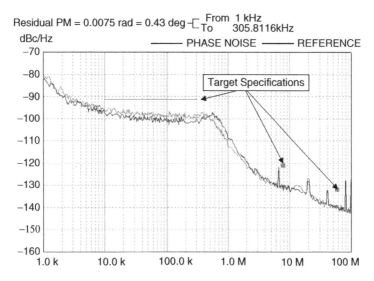

Figure 6.9 Measured phase noise of PLL of Case Study 1 at 5240 MHz. The two curves correspond to results based on two differently programmed loop filter bandwidths.

a carrier frequency of 5.24 GHz is achieved. The reference spurs are maintained to very low levels. The target specification over the bandwidth of one subcarrier is indicated by the solid line in Figure 6.9. The integrated phase modulation within the bandwidth of one subcarrier is a very important factor in the performance of an OFDM system and in this case is maintained to less than 0.45°. Also specified in the plot are the target spot specifications at 8 and 50 MHz which determine the interchannel interference performance of the system. Also shown is the capability of the PLL to adjust the loop bandwidth using JTAG programming. The loop bandwidth has greater than an octave of tuning range (not shown in Fig. 6.9).

The RF transceiver has been integrated in a 0.18-μm digital CMOS process, with a single polysilicon and five metal layers. It occupies a total area of 11.7 mm² including the padring. The chip is housed in a LPCC-48 pin package with an exposed paddle to provide good grounding. The die photo of the transceiver is shown in Figure 6.10. The chip consumes 150 mW of power in the receive mode and 380 mW of power in the transmit mode while transmitting a 15-dBm OFDM signal. The entire chip operates with a 1.8-V supply, except for the power amplifier, which operates with a 3.3-V supply. The chip passes 2.5-kV ESD testing using the human body model. The performance of the transceiver is summarized in Table 6.1. All system level measurements are referred to the chip inputs and outputs.

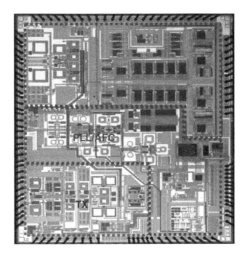

Figure 6.10 Die photo of transciever of Case Study 1.

In summary, at the time of publication of the original work (Behzad et al., 2003), this integrated circuit (IC) represented the highest performance, highest integration, smallest size, and lowest power consumption 802.11a transceiver. This chip along with the companion single-chip PHY/MAC constitutes a full 802.11a system. The transceiver achieves a 4-dB receive

Table 6.1 Summary of Performance Characteristics of Transceiver of Case Study 1

	Measured (This Work)	Unit
Frequency band	5.15–5.35	GHz
RX NF	4	dB
RX sensitivity (6 Mbps)	−93.7 ± 0.9	dBm
RX sensitivity (54 Mbps)	−73.9 ± 1.2	dBm
RX IIP3	−4.8	dBm
RX IIP2	>30	dBm
RX gain range	15–93	dB
TX power range	−30 to +18.7	dBm
TX P_{sat}	+23	dBm
TX P1dB	+19	dBm
V_{dd}	1.8	V
V_{dd} PA	3.3	V
Phase noise at 30 kHz	−100	dBc/Hz
RX power consumption	150	mW
TX power consumption	380 (15 dBm OFDM output)	mW
ESD	>±2.5 on all pins	kV
Technology	0.18 μm 1P5M CMOS	—
Die size	11.7 (including padring)	mm²

NF and a +23-dBm transmitter saturated output power. It achieves low cost through the use of a direct-conversion architecture in digital CMOS. Various integrated self-contained or system level calibration capabilities allow for high performance and high yield.

6.2 CASE STUDY 2: HIGH PERFORMANCE WLAN TRANSMITTER UTILIZING QUADRATURE AND LOFT CALIBRATION

Some of the requirements of the next generation WLAN transmitters are low transmit EVM, a low LOFT, a small I/Q imbalance, a wide gain control range, and preferably a minimum number of real-time calibrations. This case study presents a highly linear transmit mixer incorporating a wide gain control range and a *one-time* LOFT and I/Q imbalance cancellation scheme (Lee et al., 2006). First, the system architecture with the peak detector will be presented. Second, the proposed gain-controllable transconductor will be introduced. Third, the two types of LOFT and the calibration methodology to remove them as well as I/Q imbalance will be discussed. Lastly, the experimental results will be presented. Previously published methods for IQ and LOFT calibration require recalibration as a function of transmitter baseband gain control and do not distinguish between the two mechanisms of LOFT generation. Further, depending on the method of LOFT compensation, recalibration of the IQ imbalance may be required, which is highly undesirable.

Figure 6.11 shows a simplified block diagram of a direct-conversion transmitter. Each in-phase (I) and quadrature-phase (Q) baseband signal (F_{BB}) is up converted by its respective mixers. The high frequency envelope detector is driven by the PA driver output and generates a filtered and amplified baseband ripple with spectral components at F_{BB} due to LOFT and at $2F_{BB}$ due to I/Q imbalance (see Fig. 6.13 below). The idea for using an envelope detector to detect LOFT and IQ imbalance was proposed by Cavers (1997). Shown in Figure 6.11 is also the envelope detector along with the (simplified) required circuitry to amplify the detected envelope, reject the DC content, and level shift the signal to the desired common-mode level of the ADC. A wide gain control range is utilized to accommodate a wide range of possible LOFT and IQ imbalance.

In order to maximize the signal-to-offset ratio and therefore minimize LOFT in decibels relative to the carrier, it is essential to keep the baseband signal as large as possible throughout the baseband chain and any baseband gain control must be performed as close as possible to the mixer quad. A large signal in turn requires a high linearity transconductance stage for the

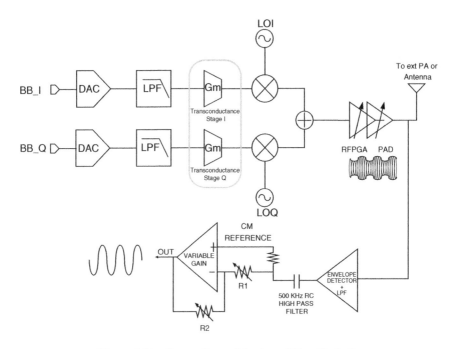

Figure 6.11   Transmitter architecture of Case Study 2.

transmitter mixer. The core of the transconductance stage is shown in Figure 6.12. The structure is similar to that proposed by Mehr et al. (1997) except that in this implementation folding is used to reduce headroom requirements. The input pair M1 and M2 is put in feedback to linearize the effective transconductance. The signal currents flow through M3 and M4 and are mirrored to M5 and M6. The effective transconductance of the circuit is $1/R_1\ (R_9/R_7)$ under high levels of degeneration. The degeneration resistors R3 to R12 reduce the circuit offsets. Note that R1 and R2 are not used as part of gain control. When R1 and R2 are changed, the overall ratio of the signal to offsets will change since the offset contribution of all the devices in the signal path following R1 and R2 will remain constant whereas all offsets prior to R1 and R2 will scale by the change in the gain.

The circled devices in Figure 6.12 constitute the proposed gain control scheme. The gates of both the shunt device, M7, and the cascode devices, M5 and M6, are tied together. To a first order, the current gain will be given by the ratio of their widths and lengths (W/Ls). Therefore, the gain control scheme is independent of process, voltage, and temperature variation and possessed a high linearity (Bult et al., 1992). A variable gain is implemented by using multiple shunt devices.

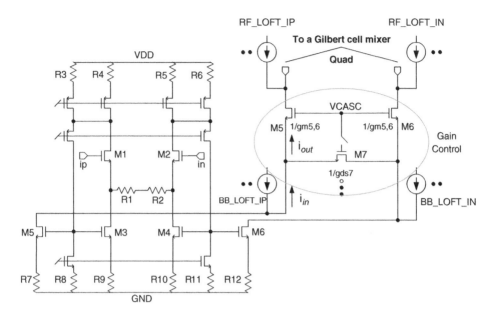

Figure 6.12 Simplified schematic of linearized transconductance stage used in up-conversion mixer of Case Study 2.

Two types of LOFT exist in a direct-conversion up-converter mixer. The first type, baseband LOFT (BB_LOFT), originates from the device offsets in the DAC, LPF, the transconductance stage, and any other baseband circuits in the signal path. The offsets will mix with the LO and generate an LO component at the output of the mixer. The second type, RF_LOFT, is a direct-coupling component either through parasitic capacitance or mutual inductance, which becomes more severe with tighter die area. The two types of offsets require a dual-cancellation scheme to cancel both components at their sources. Shown in Figure 6.12 are the two sets of currents that remove LOFT by introducing an artificial offset. BB_LOFT_IP and BB_LOFT_IN cancel all the offsets prior to the gain control stage. Note that the baseband gain control is also done at these nodes and will scale the cancellation current by the same factor, maintaining proper cancellation current without the need for a recalibration. Also note that the devices M5 and M6 are after the gain control shunt devices and could contribute a BB_LOFT at very high attenuation settings. In practice, these devices can be made sufficiently large to have minimal offsets. A second set of correction currents, RF_LOFT_IP and RF_LOFT_IN, are at the drain nodes of M5 and M6. The injection is done immediately before the switching quads of the mixer but after the gain control shunt devices so that the cancellation currents will not be affected by

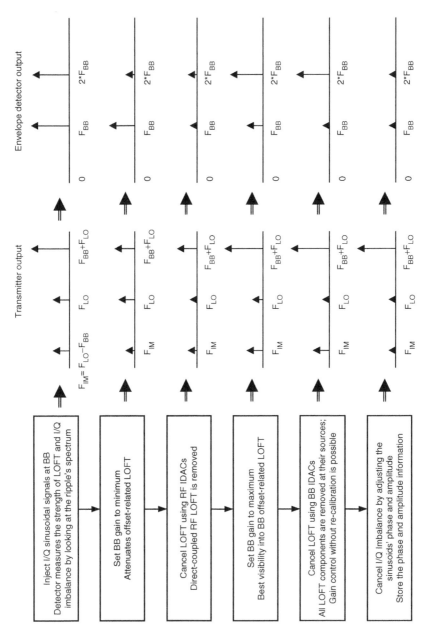

Figure 6.13 LOFT calibration algorithm used in transmitter of Case Study 2.

the gain change. Each set of correction currents are binary-weighted current DACs (IDAC).

A straightforward algorithm can be devised to separate the two LOFT components and remove LOFT as well as I/Q imbalance. Figure 6.13 shows a flow chart of the proposed algorithm. First, a set of I and Q sinusoid is injected at the baseband. Spectral analysis of the detector's output will reveal the magnitude of the LOFT (F_{LO}) and I/Q imbalance (F_{IM}). Second, the gain is set to minimum to significantly attenuate the baseband offsets. The remaining LOFT will be from RF. Third, the RF_LOFT is canceled using the RF IDAC. The digital engine could sweep through the proper IDAC codes until the F_{BB} component is minimized. Fourth, with the RF_LOFT canceled, the gain is changed to maximum. This will allow the good visibility in the event that the BB_LOFT is small. Fifth, the remaining BB_LOFT can be canceled using the BB IDAC. Note that the maximum amount of LOFT suppression will depend on the resolution of both IDACs and the achievable amount of suppression of the BB_LOFT in step 2. Lastly, keeping the gain

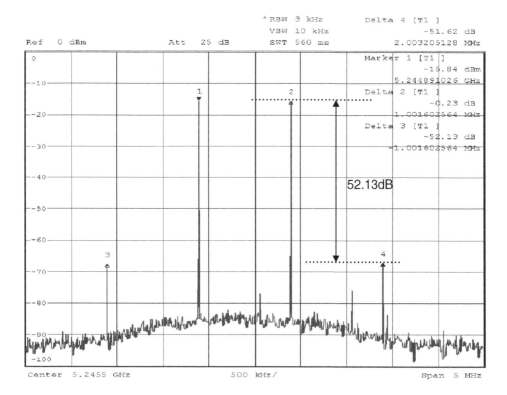

Figure 6.14 Two-tone linearity test for transmitter of Case Study 2.

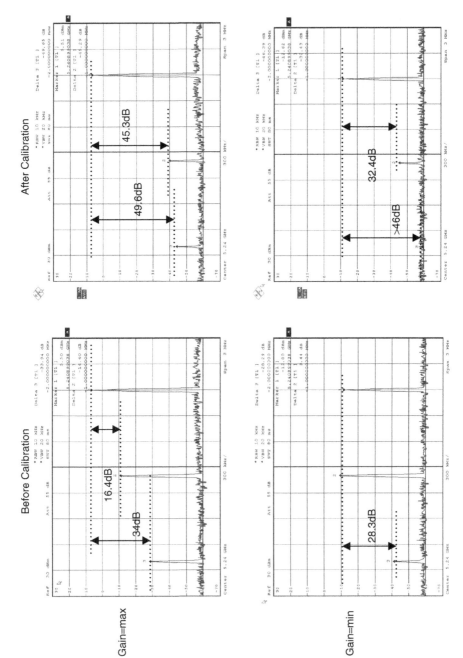

Figure 6.15 Spectrum plots of transmitter of Case Study 2 shown before and after LOFT and *IQ* balance calibration.

at maximum, $2F_{BB}$ components can be removed by adjusting the baseband signal's phase and amplitude to correct the *I/Q* imbalance. The phase and amplitude information can then be used to predistort the modulated signal in the DSP.

This chip has been fabricated in a 0.18-μm CMOS process and is in use in a multiband WLAN transceiver. Figure 6.14 shows the two-tone test result of the transmitter. To allow maximum visibility into the linearity of the transconductance stage, the transconductance stage was set to minimum gain and the following RF drivers were set to maximum gain. The small signal going into the mixer quads and the RF drivers will keep the RF stages from being the limiting components. The two-tone inputs, running at the maximum swing level of the DAC (1 V_{pp}) at 5 and 6 MHz, generate third-order intermodulation components that are more than 51 dBc down from the signal tones.

Figure 6.15 show the plots of transmitter output before and after the proposed calibration at maximum and minimum gain settings. Gain steps are

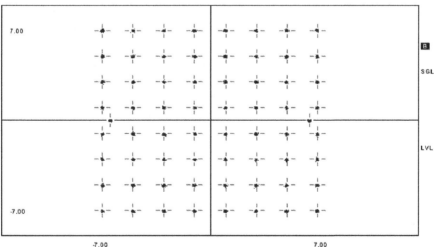

Measurement Complete

Date: 30.AUG.2005   00:11:48

Figure 6.16 Constellation diagram of entire TX chain of Case Study 2 after calibration. Output frequency is at 5.24 GHz: $P_0 = -5$ dBm, EVM <–40 dB (limited partially by lab baseband signal generator).

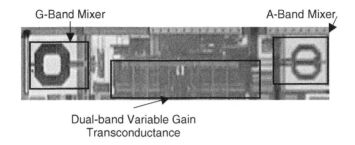

Figure 6.17 Chip microphotograph of dual-band mixer of Case Study 2.

2.5 dB and with a 17.5-dB range. The I and Q CW tones were applied at 1 MHz with the LO running at 5.24 GHz. The postcalibration LOFT is better than 32 dBc and image rejection is better than 46 dBc for all gain settings. After calibration, the residual LOFT varies slightly for different gain settings, indicating that the residual is a combination of both BB_LOFT and RF_LOFT and can be further suppressed by having a finer resolution IDACs or by using the baseband signal path DACs. The transmitter as a whole is capable of producing an output EVM of <-40 dB in the A band (Fig. 6.16) and <-41 dB in the G band. The actual EVM of the chip is somewhat better than shown in Figure 6.16, and is affected by the performance of the laboratory baseband signal generators used for this measurement.

This case study has presented a highly linear transconductance stage that incorporates gain-independent cancellation of LOFT and I/Q imbalance. The achievable performance of the scheme is limited to the resolution of the IDACs and the amount of calibration time available at the startup of the circuit. Results indicate that no other calibration other than at startup is necessary to maintain sufficient performance. The dual-band variable-gain transconductor block along with the LOFT and IQ imbalance detection circuitry occupies less than 0.1 mm^2 of die area. The chip microphotograph is shown in Figure 6.17.

CHAPTER 7

Brief Discussion of 802.11n
and Concluding Remarks

In this chapter, we will briefly review some of the highlights of the topics discussed in previous chapters, and how they relate to the latest 802.11 offering, the 802.11n. A specific 802.11n transceiver case study will be presented. We will then state some concluding remarks.

7.1 NEED FOR 802.11n

The IEEE 802.11a and g WLAN protocols provide communications data up to 54 Mbps using a 20-MHz channel bandwidth in the 2.4- and 5-GHz ISM and UNII bands, respectively. The rapid adoption of these systems in the past few years resulting from advancements in transceiver design and integration has led to a demand for more robust wireless links as well as higher data rates. These demands can be met by the adoption of MIMO techniques, the use of wider band channels, utilization of higher order constellations, and/or the use of lower coding rates for the modulation. The emerging 802.11n standard will adopt some or all of these techniques to provide a robust and high data rate wireless link. Such a standard would therefore need to utilize a high performance transceiver. In order to keep the cost down, the transceiver would need to be highly integrated and preferably implemented in CMOS technology.

As discussed in Section 1.7, multipath fading has traditionally been a fundamental barrier to achieving high data rates in wireless communication systems. Multipath is a phenomena caused by the multiple arrivals of the transmitted signal to the receiver due to reflections off of "scatterers." For traditional communication systems, multipath fading is more problematic when a direct LOS path does not exist between the transmitter and the receiver (this is commonly known as a Rayleigh fading scenario, Fig. 1.6). One problem with multipath is that it can create a selective fading response at the received signal, particularly when the bandwidth of the modulated signal is wide as compared to the "coherence bandwidth" of the channel. As shown in Figure 7.1 (Rappaport, 1996), another problem is that the received

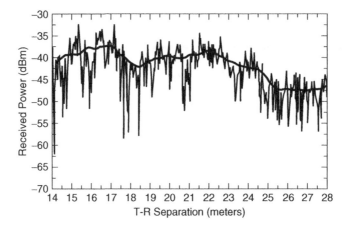

Figure 7.1   Received power variation with transmitter–receiver separation for multi-gigahertz communication system.

amplitude can vary significantly with a small change in the distance between the receiver and the transmitter.

In an all-white-Gaussian-noise (AWGN) channel, a "reasonable" SNR is sufficient for obtaining a low probability of error in the system. This is due to the fact that the only factor that can impact an error event is large additive noise. On the other hand, in a faded environment, a very large SNR would be required for a low probability of error for both coherent and noncoherent communication systems (Fig. 7.2) (Tse et al., 2005). This is due to the fact

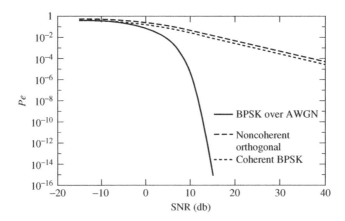

Figure 7.2   Probability of error as function of SNR in AWGN channel as well as faded channel.

that even in the presence of high SNR errors can occur due to the deep fade present in the channel.

As mentioned above, without the use of some form of "diversity," performance in a faded environment can be quite poor. One or more dimensions (*degrees of freedom*) are therefore exploited in a faded wireless system to enable diversity. Diversity can be implemented in the form of time diversity (e.g., interleaving of coded symbols), frequency diversity (e.g., intersymbol equalization, spread-spectrum techniques, OFDM), or spatial diversity (e.g., selection diversity, space–time coding). These diversity techniques can then be utilized to improve the reliability of the link.

As discussed in Chapter 1, one of the methods utilized in 802.11a/g standards for combating multipath is the use of OFDM coding. Using OFDM, the wideband modulation is subdivided in many subcarriers, each of which has a narrow bandwidth in comparison to the coherence bandwidth of a typical indoor environment. Therefore a frequency-selective fade over the wide bandwidth (Fig. 7.3) is effectively translated into flat-band fading as observed by each subcarrier. OFDM also allows for close packing of the subcarriers, due to the orthogonal nature between each two subcarriers (Figs. 1.8 and 1.9). Due to this orthogonality, the subcarriers are allowed to have overlap, since the peak magnitude of each subcarrier in frequency occurs during the null of all of the other subcarriers in an ideal environment. In reality the impairments of the transmitter, the channel, and the receiver would deteriorate the orthogonality of the subcarriers and result in reduced performance. This is one reason for the need for a high performance radio to be utilized in OFDM-based systems.

Another key feature of 802.11n is the use of multiple antennas to enhance the achievable rate as a function of the distance between the receiver and the

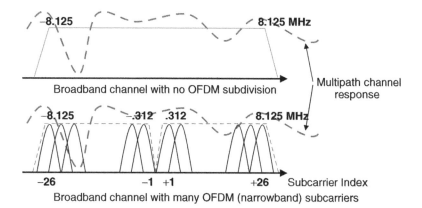

Figure 7.3 Subdivision of wideband channel by using OFDM subcarriers.

transmitter. Multiple antennas, in general, can be used for achieving power gain [e.g., receiver maximum ratio combining (MRC), transmitter beam forming (TxBF)], diversity gain (e.g., selection diversity, space–time coding), and/or degree-of-freedom gain. Whereas achieving power gain and diversity gain requires multiple antennas only on one end of the link (i.e., transmitter or receiver), achieving a degree-of-freedom gain requires a true MIMO system with multiple antennas on both the receiver side and the transmitter side (Tse et al., 2005). At the same time, a degree-of-freedom gain is the only one of the mentioned techniques that allows for an increase in the channel capacity. The 802.11n draft standard allows for the utilization of the multiple-antenna techniques mentioned in a mandatory and/or optional mode. In general, if each Rx antenna has a complete analog path to digital, RX diversity is often referred to as single-in, multi-out (SIMO). If each TX antenna has a complete analog path from digital, TX diversity is often referred to as MISO. If each RX and TX antenna has a complete analog path to and from digital, RX and TX diversity is often referred to as MIMO. These three configurations are illustrated at a high level in Figure 7.4.

The channel capacity of a legacy single-in, single-out (SISO) system is given by Shannon's famous relation

$$C_{SISO} = BW \log_2(1 + SNR)$$

where BW is the bandwidth of the channel. It can be seen that capacity increases linearly with bandwidth, which of course is a very expensive commodity. Also, the channel capacity increases only with the logarithm of the SNR. It would therefore require a very large increase in the SNR to obtain a

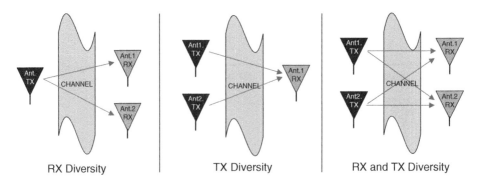

RX Diversity | TX Diversity | RX and TX Diversity

Figure 7.4 Illustration of system with receiver diversity, system with transmitter diversity, and system with receiver and transmitter diversity.

modest increase in channel capacity. On the other hand, the upper bound of the channel capacity in a MIMO environment is given by

$$C_{\text{MIMO}} = \min(n, m) \text{ BW } \log_2(1 + \text{SNR})$$

where n is the number transmitter antennas and m is the number of receiver antennas. The channel capacity can now be increased in a linear fashion with the minimum of the transmitter and receiver antennas. Although this relation provides for an upper bound of the capacity, in practice, numbers very close to the upper bound are achievable.

Many analog techniques have been utilized in the past to effectively combat multipath fading in a wireless system. These include simple antenna diversity selection or more sophisticated analog maximum-ratio-combining techniques using phase shifters or programmable delay elements. However, a "true MIMO" system (one where each receiver and transmitter antenna is connected to the baseband processor through a dedicated analog chain) provides for many advantages (Fig. 7.5). For example, phase shifting, combining, and beam forming can be performed easily at digital baseband on a *persubcarrier* basis in OFDM systems. Further, such a true MIMO system allows for "spatial multiplexing," thereby increasing the channel capacity dramatically, as discussed earlier. On the other hand, such true MIMO systems consume more power and area than analog-only multiantenna solutions.

The draft 802.11n standard combines the capabilities offered by OFDM coding along with those offered by multiple antennas. As a result, a MIMO–OFDM system is constructed which offers diversity gain as well as

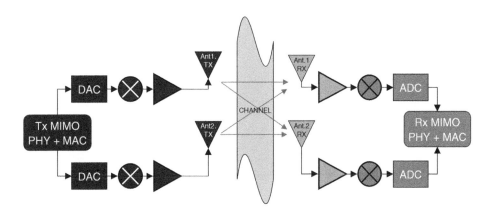

Figure 7.5 "True" MIMO system in 2 × 2 configuration.

power gain and utilizes a rich scattering environment to offer a significant increase in the channel capacity. In such a MIMO–OFDM system (Fig. 7.6), a high speed data stream is split into multiple slower streams and the independent data streams are sent on each transmit antenna at the same time and at the same frequency band using a space–time processor. The received data streams which have been subject to the scattering channel are received on the multiple receive antennas, separated out by the space–time processor on the receive side, decoded, and then combined to regenerate the high speed data stream.

The draft 802.11n standard offers many features and capabilities. These include spatial division multiplexing (SDM) through the use of MIMO–OFDM, bandwidth expansion, higher rate binary convolutional codes, new frame formats, reduced interframe spacing, short guard intervals, space–time block codes, transmit beam forming, low density parity check codes, new modulation and coding schemes (MCS), and a variety of aggregation techniques. The radio presented in this case study is capable of supporting all of these functions (Behzad et al., 2007a).

7.2 802.11a/b/g/n MIMO TRANSCEIVER

A single-chip fully integrated multiband direct-conversion CMOS MIMO transceiver targeted for WLAN applications is described in this case study

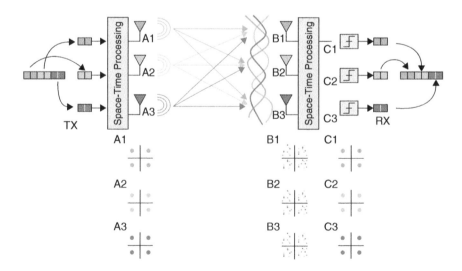

Figure 7.6 MIMO–OFDM system concept.

(Behzad et al., 2007b). This transceiver is capable of satisfying the requirements of the Enhanced Wireless Consortium (EWC) and the draft version of the 802.11n as it stands today. The transceiver presented here is implemented in a 2 × 2 format (i.e., two receivers and two transmitters) and is capable of supporting data rates of over 250 Mbps. Each receiver and transmitter is capable of operating at the 2.4- to 2.5-GHz as well as the 4.9- to 5.9-GHz bands. Additionally each receiver and transmitter is capable of working with modulated signals with (RF) modulation bandwidths of 10 to 40 MHz. The receivers and transmitters are fed by a shared PLL through LO generation and distribution circuitry. The receivers and transmitters achieve an EVM of better than –41 dB (0.9%) and –40 dB (1.0%) operating in legacy G and A modes, respectively (results limited by lab OFDM signal generators). From a 1.8-V supply and with both cores operating, the chip consumes 275 mA in the RX mode and 280 mA in the TX mode. As compared to the previously published MIMO transceivers for WLAN (Chien et al., 2006; Palaskas et al., 2006; Rahn et al., 2005), this transceiver is implemented in a low cost 0.18-μm CMOS technology, covers both the G band as well as the entire worldwide A band, is significantly smaller in area, is capable of operating over RF channel bandwidths of 10, 20, or 40 MHz, and achieves much lower NF, PN, and EVM. Furthermore, for the first time, transceiver performance under various draft 802.11n MCSs are presented and PHY rates of > 270 Mbps are shown.

7.2.1 Architecture and Circuit Implementation

The full 2 × 2 MIMO transceiver is composed of two multiband RX chains, two multiband TX chains, a PLL and LO generation and distribution section, various calibration blocks, a digital control section, and some other miscellaneous circuits. The high level block diagram is shown in Figure 7.7. Each subblock is described in the following sections.

7.2.1.1 Receiver

A block diagram of each receiver slice is shown in Figure 7.8a. (Note that for simplicity all block diagrams are shown with single-ended traces, but the actual transceiver is designed with fully differential signaling throughout the chip.) The 5- or 2-GHz signal is received and amplified by the appropriate differential LNA, then directly down converted by the quadrature mixers associated with that band. The down-converted signal is then applied to the first high pass VGA (HPVGA1) where the signal is amplified and the DC offsets associated with the self-mixing of the mixers and device mismatches are rejected. The signal is then filtered by the fourth-order Butterworth fil-

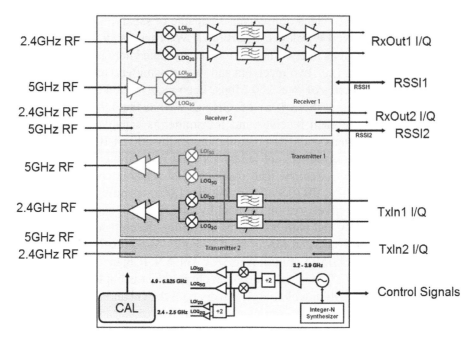

Figure 7.7 Transceiver high level block diagram.

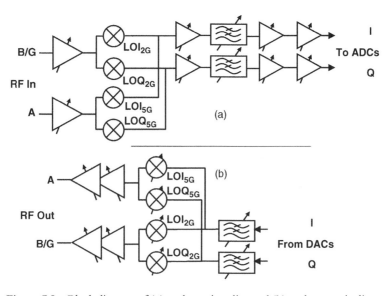

Figure 7.8 Block diagram of (**a**) each receive slice and (**b**) each transmit slice.

ters which act to reject any interferers. The signal is then further amplified by HPVGA2 and HPVGA3, each with their own DC offset cancellation loops. The resultant I and Q outputs are then buffered and sent to the I and Q ADCs on the companion PHY+MAC chip. Each HPVGA has a programmable gain of 0 to 30 dB in 3-dB steps. The corner frequency of the HPFs in the HPVGAs are calibrated using an on-chip RC calibration loop. Further, the HPF corners are programmable over a wide range to satisfy the integrity of the lower index OFDM subcarriers during the payload while allowing for fast settling during the preamble. The corner frequency of the LPF is also calibrated to the desired bandwidth (5, 10, or 20 MHz) to ensure the proper system operation in the presence of large adjacent channel interferers. Two wideband RSSI signals as well as one narrowband RSSI ensure the proper operation of the system over a very wide dynamic range and in the presence of large (CCK, OFDM, BT, CW, etc.) interferers. Each receiver slice is capable of >100 dB of total gain, >100 dB of gain control range, an input IIP3 of +5 dBm with the RF front end at low gain, and a NF of 4 dB with the RF front end at maximum gain.

7.2.1.2   Transmitter

Figure 7.8b displays the block diagram of each transmitter slice. Received quadrature signals from the DACs are applied to the programmable bandwidth and gain low pass filters. The output of the LPFs are then applied to the appropriate up-conversion quadrature mixers which directly convert the baseband signal to the desired RF band. The up-conversion mixers (with associated transconductors) are designed for high linearity and low LOFT over a wide gain control range (Lee et al., 2006). The RF signal is then amplified through two stages of programmable gain. The final gain stage is capable of driving a 50-Ω load through a balun and is internally matched to a 100 Ω differential. The RF gain stages are designed such that their power consumption is reduced at lower gain settings.

The core of the transconductance stage for this chip was shown in Figure 6.12 and is repeated for convenience in Figure 7.9. The input pair, M1 and M2, is put in feedback to linearize the effective transconductance. The signal currents flow through M3 and M4 and are mirrored to M5 and M6. The effective transconductance of the circuit is $1/R_1 \times R_9/R_7$ under high levels of degeneration. The degeneration resistors R3 to R12 are used to reduce the device offsets. Note that R1 and R2 are not used as part of gain control. When R1 and R2 are changed, the overall ratio of the signal to offsets will change since the offset contribution of all the devices in the signal path following R1 and R2 will remain constant whereas all offsets prior to R1 and R2 will scale by the change in the gain. The circled devices in Figure 7.9

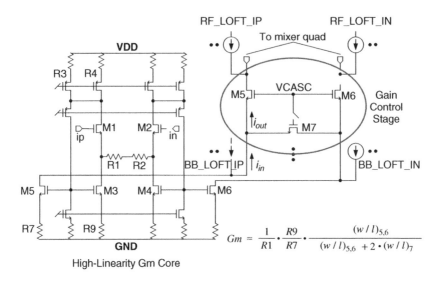

Figure 7.9 High linearity TX mixer transconductor and gain control stage. LOFT cancellation current sources also shown.

constitute the proposed gain control scheme. The gates of both the shunt device, M7, and the cascade devices, M5 and M6, are tied together. To the first order, the current gain will be given by the ratio of their W/L's. Therefore, the gain control scheme is independent of process, voltage, and temperature variation and possesses a high linearity. A variable gain is implemented by using multiple shunt devices. The four current sources that are shown are utilized to cancel LOFT due to DC offsets as well as direct RF coupling during startup calibration. No calibration is required as the gain of the block is changed. The calibration algorithm was discussed in Section 6.2.

Figure 7.10 shows the simplified schematic of the PA driver. A three-stage transconductance linearization is used to improve linearity over a wide range of inputs (Behzad et al., 2004). The first stage, biased well in the class A region, supplies the majority of the gain but is only linear over a small range of inputs. The second and third stages, biased at class AB and class B, respectively, contribute more to the gain as input increases and counteract the gain roll-off in the first stage. The class AB nature of the driver saves valuable DC when the circuit is not transmitting large signals. The obtained transconductance curves for each subsection of the PA driver as well as the overall resultant transconductance are shown in Figure 7.11. The transmitter has an output P1dB of +14 dBm in the A band and an output P1dB of +16 dBm in the G band.

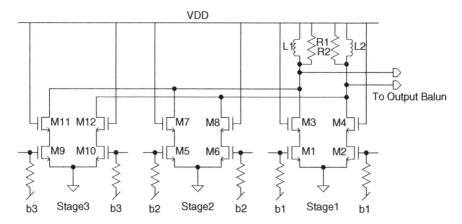

Figure 7.10   A-band PA driver transconductance linearization.

7.2.1.3   PLL and LO Generation

A shared integer-N PLL is used to synthesize the proper LO (Fig. 7.12). The on-chip crystal oscillator whose divided down output provides the reference to the PFD is capable of operating with a 20-, 40-, or 80-MHz crystal. In order to minimize the impact of the VCO noise, The PFD comparison frequency is chosen to be as high as possible to allow wide PLL bandwidth. Further, in order to minimize the phase noise arising from the loop filter and charge pump, a low programmable KVCO (30 MHz/V at 3.5 GHz, typical)

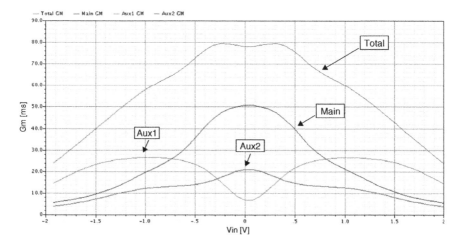

Figure 7.11   Transconductance curves for A-band PA driver stage. The main, aux1, aux2, and total transconductance curves are shown.

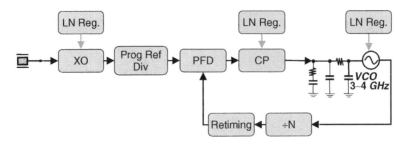

Figure 7.12 Block diagram of shared integer-N PLL utilized on MIMO transceiver. Very low noise regulators are utilized on sensitive blocks to obtain a very low phase noise design.

and high charge pump current are used. The PLL utilizes a single VCO with a wide tuning range to cover both 802.11 frequency bands. A 9-bit high accuracy VCO calibration is implemented to choose the best VCO subband for the desired channel. Every block in the synthesizer, including the crystal oscillator and the bias and supply generation circuitry for the PLL, are designed, simulated, and optimized for optimal noise performance. Three very low noise regulators are integrated on-chip and provide a low noise supply voltage to the crystal oscillator, the charge pump, as well as the VCO. All of the components of the PLL other than a large capacitor in the loop filter are integrated on-chip. The block-by-block noise simulation results are then used to obtain an overall closed-loop PLL phase noise response for the various channels. The simulation results for 5420 MHz are shown in Figure 7.13. These simulated results match the measured results very closely.

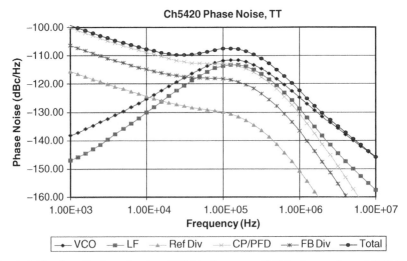

Figure 7.13 Contribution of various PLL blocks to overall PLL phase noise at 5420 MHz.

In order to avoid any pulling effects by any of the transmitters on the VCO, the VCO operates at two-thirds the channel frequency for the 802.11a band and at three-fourths the channel frequency for the 802.11b/g bands (Fig. 7.14). A differential and quadrature LO signal is required for the operation of each RX and TX mixer at the 2.4- or 5-GHz bands. Proper LO buffering and distribution are therefore essential for the robust operation of the system over PTV as well as for reducing power consumption. Therefore, the LO distributions channels were accounted for in the very early stages of the floor planning of the chip. Each local buffer is designed to be able to drive the local mixer as well as the routing and the next buffer stage (Fig. 7.14). As such the LO distribution of the chip is scalable to larger integrated $M \times N$ MIMO systems.

7.2.1.4 Calibration Techniques

The transceiver utilizes extensive self-contained and/or DSP-assisted autocalibration circuitry. These calibrations include independent and multiphase LOFT calibration (RF coupling and DC offsets) on each transmit core; independent quadrature (phase and amplitude) calibration on each receive and transmit core; resistor calibration for bias current generation blocks; LPF and HPF corner calibration on filters; multiphase VCO calibration (openloop, closed-loop, and refresh mode); and VCO gain calibration and transmitter power calibration. Additionally a high accuracy temperature sensor allows for temperature-based calibrations if necessary. Some of these calibrations may have been desirable but not absolutely necessary for some legacy WLAN ICs but are essential for obtaining the performance at the lev-

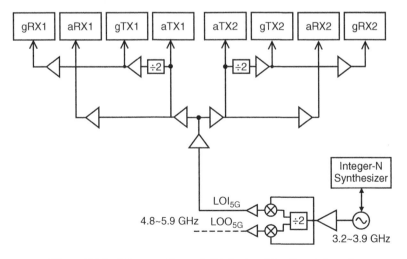

Figure 7.14 LO generation and distribution (Q path not shown).

els that are required by this RFIC. These calibrations ensure optimal operation over process, temperature, and voltage as well as a high yielding part.

The transmitter LOFT and *IQ* calibration technique has been discussed in Section 6.2 and will not be repeated here.

In order to obtain optimal receiver performance, especially for the 5-GHz band, RX *IQ* calibration is required. One method would be to utilize the calibrated transmitter to transmit a single-sideband test tone and couple this test tone to the receiver. Then the down-converted *I* and *Q* signals at the receiver baseband can be used to calibrate the RX *I* and *Q* in digital baseband. In the basic implementation of such a scheme in a MIMO radio, the TX1 output would be used to calibrate RX1, and the TX2 output would be use to calibrate RX2 (Fig. 7.15). The issue with this approach is that the loading of BUF21 is different during calibration and during normal operation. The same can be said about the loading of BUF22. With this scheme, therefore, postcalibration image rejection can be limited to –35 dBc. In the current chip, an alternative cross-core calibration method has been utilized (Fig. 7.16). In this approach the output of TX2 is coupled to the input of RX1 during calibration. Therefore the loading of BUFF22 is constant during calibration and normal operation. The procedure is repeated for RX2 once RX1 is calibrated. With this approach, postcalibration image rejection is better than –50 dBc.

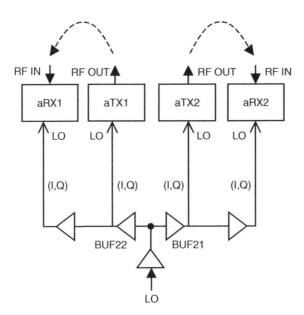

Figure 7.15 RX *IQ* calibration—same core method.

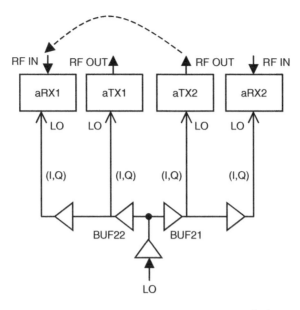

Figure 7.16 RX *IQ* calibration—cross-core method.

7.2.2 Packaging Issues

In addition to the challenges associated with designing a single transceiver for high EVM performance, there are many other potential complications that need to be addressed in order to be able to integrate multiple such transceivers on the same Si substrate. For example, this chip includes 16 RF input/outputs (IOs), 16 baseband analog signal path IOs, several other analog IOs (RSSIs, TSSIs, temperature sensor, etc.), and tens of high speed digital IOs. Unfortunately such a chip cannot be integrated in a "RF-friendly" package such as a leadless package chip carrier (LPCC), and larger, more complex packages are required. In this case a 165-ball fine-pitch ball-grid array (fpBGA) package is utilized. Full three-dimensional electromagnetic simulations were conducted on the package to ensure proper operation and isolation of the numerous RF sections of the chip (Fig. 7.17). In order to ensure optimal performance, the chip design, chip floor planning, package design, and many aspects of the board design were conducted simultaneously and further complicated the overall design task.

7.2.3 Measurement Results

The phase noise of the PLL for the 5.240-GHz A bands is shown in Figure 7.18. The phase noise plot is obtained with the PLL and crystal oscillator

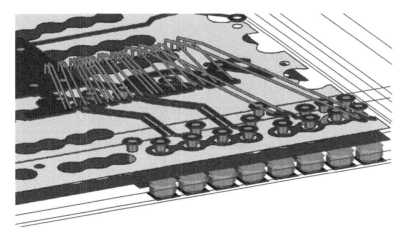

Figure 7.17 Design of fpBGA package substrate. The RF section of one core shown.

(XO) voltage supply being provided by the integrated ultralow noise regulators and with a 20-MHz crystal used as the reference. As shown in the EVM–subcarrier plot of Figure 7.19, operating in the legacy 802.11g mode, each transmitter achieves an EVM of –41 dB while transmitting –2 dBm. As shown in the constellation diagram of Figure 7.20, in the legacy 802.11a mode, each transmitter achieves an EVM of –40 dBm at –5 dBm TX power. The constellation diagram at the nominal received power level as well as at

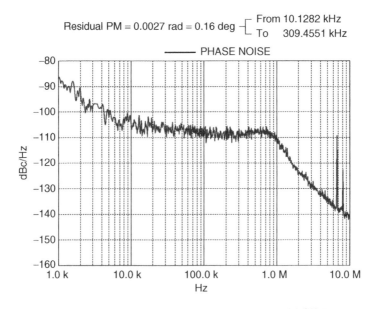

Figure 7.18 Measured PN at transmitter output (5.24 GHz).

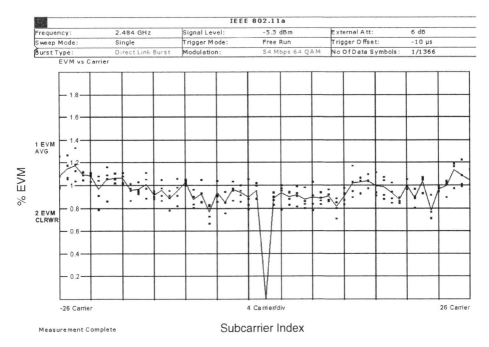

IEEE 802.11a						
Frequency:	2.484 GHz	Signal Level:	-5.3 dBm	External Att:	6 dB	
Sweep Mode:	Single	Trigger Mode:	Free Run	Trigger Offset:	-10 µs	
Burst Type:	Direct Link Burst	Modulation:	54 Mbps 64 QAM	No Of Data Symbols:	1/1366	

Figure 7.19 Measured TX EVM ($P_0 = -2$ dBm; -41 dB at 2.484 GHz).

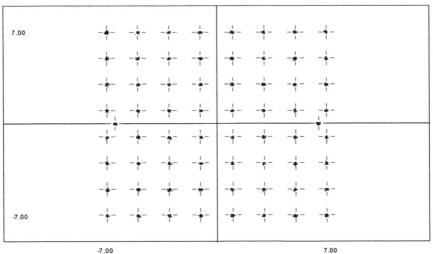

IEEE 802.11a						
Frequency:	5.24 GHz	Signal Level:	-9.8 dBm	External Att:	6 dB	
Sweep Mode:	Single	Trigger Mode:	Free Run	Trigger Offset:	-10 µs	
Burst Type:	Direct Link Burst	Modulation:	54 Mbps 64 QAM	No Of Data Symbols:	1/1366	

Figure 7.20 Measured TX constellation diagram ($P_0 = -5$ dBm; -40 dB at 5.2 GHz).

the sensitivity level of the receiver for the legacy 802.11g 54-Mbps packets is shown in Figure 7.21 (2.412 GHz). Chip-referred sensitivity level of –78 dBm (at each antenna) is achieved in the MRC mode. Under similar conditions the sensitivity for a 802.11a signal at 5.24 GHz is –79 dBm. In the draft 802.11n MCS15, 40-MHz mode, AWGN channel, with standard guard interval (GI) (270 Mbps PHY rate), a chip-referred sensitivity of –72 dBm is achieved. These EVM and sensitivity results are an indication of the excellent analog characteristics of the transceiver, such as phase noise, *IQ* balance, linearity, and noise figure. All RX and TX EVM measurements are limited and affected by the EVM of the lab signal generation and VSA equipment, which are only a few decibels better in performance than the device under test.

Figure 7.22a displays the average effective throughput of the system in a Chariot test. The chipset is set to the MCS15 mode, with channel bandwidth of 40 MHz, and standard GI. In this setting the PHY rate is 270 Mbps. The 2442-MHz G band channel is utilized for this test. A throughput of >200 Mbps is achieved. This measurement is performed in an unconstrained system where the central processing unit or transmission control protocol would not limit the throughput of the system. Figure 7.22b displays short-

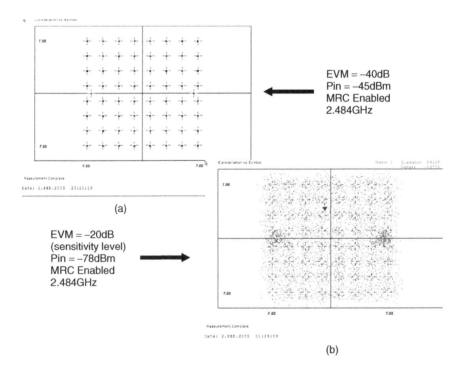

(a)

EVM = –40dB
Pin = –45dBm
MRC Enabled
2.484GHz

EVM = –20dB
(sensitivity level)
Pin = –78dBm
MRC Enabled
2.484GHz

(b)

Figure 7.21   Measured RX constellation diagram.

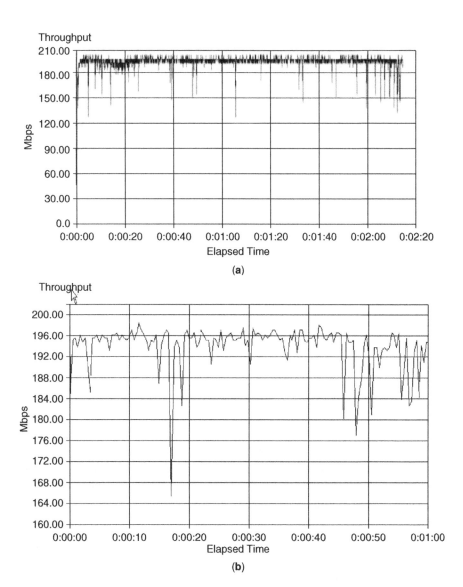

Figure 7.22   Real-world system performance/throughput. (**a**) Cable test, 2 × 2 system, throughput > 200 Mbps, PHY rate = 270 Mbps, 2.442-GHz channel. (**b**) Close range (10 ft) over the air at 5.24 GHz, 2 × 2 system, max throughput > 193 Mbps.

range over-the-air testing at 5.24 GHz where a maximum throughput of >193 Mbps is achieved.

Figure 7.23 displays the throughput versus range for a combined line-of-sight and non-line-of-sight channel. This test is performed in a 2 × 2 configuration with the radio set to the 2.442-GHz channel. The results are

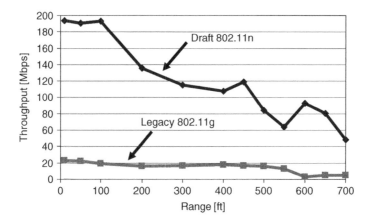

Figure 7.23 Effective system throughput versus range for system utilizing this radio and comparison against a legacy 802.11g system.

compared against a legacy 802.11g system. The advantages of the draft 802.11n MIMO system as compared to a legacy system are apparent at both the short and long range.

As mentioned earlier, small antenna displacements can result in large channel capacity variations. When three antennas are available in the system, the best two antennas can be selected by utilizing a spatial probing algorithm. This algorithm is operated as an extension of the rate selection algorithm. The radio presented here can be utilized in this 3×3 dynamic antenna selection system when three antennas are available. This dynamic antenna selection algorithm maximizes throughput at a particular location, maximizes the throughput under worst channel conditions, and maximizes the average throughput of the system. As a result, the system is able to maintain consistent and high throughput over time.

The transceiver performance summary is presented in Table 7.1.

7.2.4 MIMO Case Study Conclusion

A very high performance, low power, small area, multiband draft 802.11n transceiver has been implemented in a digital CMOS process and presented as a case study here. The transceiver is capable of supporting PHY rates of >270 Mbps and an effective throughput of >190 Mbps in real-world over-the-air testing. Various calibration techniques have been utilized to enable high performance, low power, and/or high production yield.

The chip die photo is shown in Figure 7.24b. The chip integrates the equivalent of approximately 100 single-ended inductors. The IC occupies a total die area of 18 mm² in a digital 0.18-μm CMOS process and is packaged in a 165-ball fpBGA package. The chip is designed such that it can be utilized in a mul-

Table 7.1 Transceiver Performance Summary

	Measured	Spec.	Unit
Frequency band (A/G)	2.4–2.5/4.8–5.9		GHz
RX max. gain (A/G)	>100/>100		dB
RX min. gain (A/G)	5/5		dB
RX NF at max. RF gain (A/G)	4/4.5		dB
RX IIP3 at min. RF gain (A/G)	+5/+6		dBm
RX IIP3 at max. RF gain (A/G)	−12/−10		dBm
RX-RX isolation (A/G)	>60/>60	>30	dB
RX LPF BW	5/10/20		MHz
TX P1dB (A/G)	+14/+16		dBm
TX EVM (A/G)	<−40/<−41	<−30	dB
TX-TX isolation (A/G)	>35/>43	>15	dB
In-band PN at 150 kHz effect (A/G)	−108/<−109	<−100	dBc/Hz
Legacy 54 Mbps, chip-referred MRC-enabled sensitivity (A/G)	−79/−78	−65	dBm
EWC MCS15, 40-MHz channel, std. GI, 270 Mbps, chip-referred sensitivity	−72/−72		dBm
V_{dd}	1.8		V
RX mode total current consumption	275		mA
TX mode total current consumption (both cores active)	280		mA
Technology	0.18 μm CMOS		
Die size	18		mm²
ESD HBM performance	> ±2		kV

titransceiver configuration to build a larger MIMO system (e.g., 4 × 4). The chip passes better than ±2-kV ESD performance on all pins. As a comparison, the die photo of a first-generation 802.11a transceiver implemented in the same process technology is shown to scale in Figure 7.24 (Behzad et al., 2004b). This 802.11a transceiver occupies 12 mm². The fact that a dual-core, dual-band, triple-mode transceiver can be implemented in only 1.5 times the area of a single-core, single-band transceiver of only a few years ago is testimony to the advancements in architectures and circuit techniques in the past few years.

7.3 CONCLUDING REMARKS

In conclusion we have briefly touched upon the various topics in the area of WLAN radio design. Complex trade-offs between the following topics were discussed: the various flavors of WLAN, the radio architectures, the analog impairments, choice of process technology, and various forms of autocalibration. Two specific case studies were presented.

All indications at this point are that a trend toward higher data rates and higher spectral efficiencies on the high end of the market will continue.

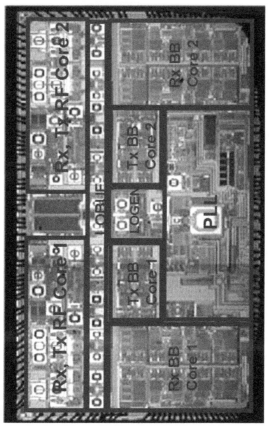

(b)

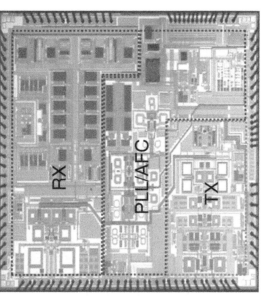

(a)

Figure 7.24 (a) Die microphotograph of SISO 802.11a transceiver (Behzad et al., 2003). (b) Die microphotograph of this dual-band MIMO transceiver.

Trends toward lower cost, smaller size and lower power consumption, and lower end of the market will also continue. Further, the trend of integration into embedded applications is emerging and will drive the growth of WLAN chip sales in the next several years.

Radios will become cheaper, smaller, and better by utilizing innovative systems and circuit techniques as well as taking advantage of the power of DSP for self-calibration. Multifunction and multistandard transceivers will become more commonplace and researchers will continue to work toward "software-definable" radios.

Figure 7.25 displays two sides of one module with a single-chip Bluetooth BCM2035 mounted on the top side of the printed circuit board (PCB)

Figure 7.25 Two sides of one module with single-chip Bluetooth BCM2035 (top picture) on one side and single-chip WLAN BCM4317 (bottom picture) on the other side.

and a single-chip wireless LAN BCM4317 on the bottom side of the PCB. As can be seen, the packaged IC dimensions are small, the external components are few, and the whole system has been integrated in a very small footprint. BCM2035 integrates the Bluetooth PHY, MAC, and transceiver. BCM4317 integrates a WLAN PHY, MAC, transceiver, as well as the power amplifier and the TR and diversity switches.

A more recent example of the advancement of the art is the BCM4325 which integrates full 802.11a/b/g functionality, Bluetooth enhanced data rate (EDR), as well as an FM receiver in a single 65-nm digital CMOS IC.

In order to satisfy the needs of the communication gear of tomorrow, innovations in process technology, IC design, system design, MAC design, packaging technology, PCB technology, and many other areas of engineering will have to continue. These innovations will lead us to a ubiquitous connected world.

About the Author

Arya Behzad, Broadcom Distinguished Engineer, is a director of engineering working on radios for current and future generation wireless products and product-line manager for all wireless LAN radio products. Also at Broadcom, he is recognized as one of the most influential contributors to CMOS RF R&D efforts and his product shipments have surpassed the 200-million unit mark. Mr. Behzad has more than 100 patents issued and pending, as well as many publications in the areas of precision analog circuits, cellular transceivers, integrated tuners, gigabit Ethernet, and wireless LANs. He has taught courses and presented technical seminars at various conferences and at several universities. This book and his *IEEE Expert Now* course on wireless LAN radio design are both derived from his popular IEEE ISSCC course on the same topic. Mr. Behzad is serving a sixth year as a member of the IEEE International Solid State Circuits Conference Wireless Technical Committee. In the past, he has served as a guest editor of the *IEEE Journal of Solid State Circuits* and is currently an associate editor of the journal. He is a Senior Member of the IEEE. In 1994, he earned a Master of Science degree in Electrical Engineering from the University of California at Berkeley.

Annotated Bibliography

1. A. Behzad, Wireless LAN Radios: System Definition to Transistor Design, IEEE Expert Now interactive course, http://www.ieee.org/web/education/Expert_Now_IEEE/modules.html.

   An interactive course on the material covered in this book.

2. http://www.ieee802.org/.

   The official link to the IEEE 802 LAN/MAN standard committee.

3. http://grouper.ieee.org/groups/802/11/QuickGuide_IEEE_802_WG_and_Activities.htm.

   The link to the quick guide into the IEEE 802.11 working group and activities.

4. A Montalvo, "Highly Integrated RF & Wireless Transceivers," IEEE International Solid-State Circuits Conference Tutorial, San Francisco, 2003.

   This tutorial provides an introduction to the challenges associated with highly integrated RF and wireless transceivers. Both system level and circuit level challenges are addressed. The focus is on transceivers for high data rate systems such as EDGE, WCDMA, and 802.11, which require linearity in both the TX and RX paths. Some of the topics that will be addressed are architectures such as direct-conversion RX and TX which enable elimination of external components, circuit level issues associated with implementing these architectures, the impact of finite isolation, and manufacturability.

5. T. Tuttle, "Introduction to Wireless Receiver Design," IEEE International Solid-State Circuits Conference Tutorial, San Francisco, 2002.

   An introduction to integrated receivers focuses on performance requirements for GSM cellular handset applications: (1) overview of radio standards for sensitivity, blocking, AM suppression, and intermodulation; (2) comparison of heterodyne, direct-conversion, and low IF receiver architectures; (3) discussion of design specifications and trade-offs.

6. D. Shoemaker et al., "Wireless LAN: Architecture and Design," IEEE International Solid-State Circuits Conference Tutorial, San Francisco, 2003.

   An overview of competing wireless LAN architectures in the 2.4- and 5.0-GHz spectra is presented. RF and analog requirements will be discussed. A detailed discussion of competing modem algorithms (OFDM, QAM, QPSK, CCK, etc.) will be presented, with advantages and disadvantages identified for each. Filtering requirements and some aspects of the software protocol stack will be compared. The tutorial will end with some implementation ideas for the harder issues and a performance analysis for each system.

7. B. Cutler, "Effects of Physical Layer Impairments on OFDM Systems," *RF Design,* May 2002, pp. 36–44.

   The effects of common signal impairments using single-carrier modulation formats are generally well understood by system designers. The effects of these same impairments on an OFDM signal, however, can be quite different and are discussed in this article.

8. R. VanNee and R. Prasad, *OFDM for Wireless Mulimedia Communications*, Artech House, 2000.

   A book on the digital PHY section of OFDM systems. It offers a detailed treatment of practical OFDM system concepts and their applications to practical communication concepts.

9. I. Bouras et al., "A Digitally Calibrated 5.15GHz–5.825GHz Transceiver for 802.11a Wireless LANs in 0.18 μm CMOS," *ISSCC, 2003*.

   This paper presents a single-chip 5-GHz fully integrated direct-conversion transceiver for IEEE 802.11a WLAN systems manufactured in 0.18-μm CMOS. The IC features a system architecture which takes advantage of the computing resources of the digital companion chip in order to eliminate I/Q mismatch and achieve accurately matched baseband filters. The integrated VCO and synthesizer achieve an integrated phase noise of less than 0.8° RMS. The receiver has an overall noise figure of 5.2 dB and achieves sensitivity of –75 dBm at 54-Mb/s operation, both referred to the IC input. The transmit error vector magnitude is –33 dB at –5 dBm output power from the integrated power amplifier–driver amplifier. The transceiver occupies an area of 18.5 mm².

10. M. Zargari, "Challenges in the Design of a CMOS RF Transceiver for IEEE 802.11a Wireless LAN," VLSI Symposium Short Course, Hawaii, 2003.

    This paper presents the challenges involved in the design of integrated IEEE 802.11 wireless LAN transceivers with focus on radio architecture and circuit implementation. In particular, examples of critical blocks in receiver and transmitter are discussed.

11. D. Su et al., "A 5GHz CMOS Transceiver for 802.11a Wireless LAN," IEEE International Solid-State Circuits Conference Tutorial, San Francisco, 2002.

    A 5-GHz transceiver comprising the RF and analog circuits of an IEEE 802.11a-compliant WLAN has been integrated in a 0.25-μm CMOS technology. The IC has 22 dBm maximum transmitted power, 8 dB overall receive chain noise figure, and –112 dBc/Hz synthesizer phase noise at 1-MHz frequency offset.

12. A. Behzad et al., "A Direct-Conversion CMOS Transceiver with Automatic Frequency Control for IEEE 802.11a Wireless LAN," IEEE International Solid-State Circuits Conference Tutorial, San Francisco, 2003.

    A 11.7-mm² 5-GHz direct-conversion 0.18-μm CMOS transceiver achieves a sensitivity of –93 dBm, a system NF of 4.5 dB (high gain), and IIP3 of –4.8 dBm (low gain). Dissipation is 150 mW in RX mode and 380 mW while transmitting 15-dBm OFDM signal.

13. T. Schwanenberger et al., "A Multi Standard Single-Chip Transceiver Covering 5.15 to 5.85GHz," IEEE International Solid-State Circuits Conference Tutorial, San Francisco, 2003.

    This transceiver achieves a transmit 1-dB output compression point of +15 dBm, and the overall receiver noise figure is 5 dB. A power gain range of >45 dB/65 dB for transmit/receive and a PLL synthesizer frequency range of 4.9 to 5.85 GHz with –79 dBc/Hz phase noise at 10-kHz offset have been measured. The IC is realized in 0.5-μm SiGe BICMOS technology and occupies 17 mm².

14. J. Troychak, "The Design and Verification of IEEE 802.11a 5GHz Wireless LAN System," Technical Note, Agilent-EESOF.

    This article reviews the IEEE 802.11a physical layer, discuses system level design considerations, and describes an integrated software/hardware design flow for IC design and verification.

15. "Making 802.11g Transmitter Measurements," Application Note 1380-4, Agilent.

    This application note outlines the transmitter measurements required for 802.11g and examines how they relate to measurements needed for 802.11a and 802.11b testing.

16. "RF Testing of Wireless LAN Products," Application Note 1380-1, Agilent.

    This application note looks at the modulation technology behind several WLAN standards and the measurement techniques that can be used to troubleshoot and quantify their RF performance. The emphasis will be on 802.11b, 802.11a, and HIPERLAN Type 1 and Type 2. The principal focus of this document is the physical RF layer of WLAN signals, as opposed to the MAC layer or higher layers of a WLAN signal. This includes time, frequency, and modulation domain analysis and troubleshooting as well as the basic modulation theory behind these standards.

17. B. Come, "Impact of Front-End Non-Idealities on Bit Error Rate Performance of WLAN-OFDM Transceivers," paper presented at the Radio and Wireless Conference, Denver, 2000.

    Given the high data rates with required low bit error rates of 802.11a and given the nature of the OFDM signal, a conservative analysis of the front-end requirements lead to severe, overdimensioned specifications. Such a design would never meet this market, by necessity low cost and low power. To extract more optimal front-end specifications, this paper assesses the BER performance of the complete WLAN–OFDM link. As a result, it first shows that the transmitted symbols' word length can be restricted to 8 bits and the normalized crest factor digitally limited at baseband to 4. Then it shows that the power amplifier can operate with only 5.4 dB backoff between the average input power and the input-referred P1dB. Finally, we quantify in terms of implementation loss the influence of the I/Q imbalance and of the frequency synthesizer phase noise.

18. A. Abidi, "Direct Conversion Radio Transceivers for Digital Communication," *JSSC,* Dec. 1995.

    This paper briefly covers case studies in the use of direct-conversion receivers and transmitters and summarizes some of the key problems in their implementations. Solutions to these problems arise not only from more appropriate circuit design but also from exploiting system characteristics, such as the modulation format in the system.

19. A. Abidi, "CMOS Wireless Transceivers: The New Wave," *IEEE Communications Magazine,* Aug. 1999, pp. 119–124.

    This article reviews some of the RF CMOS circuit design techniques and shows how an understanding of the strengths and weaknesses of these circuits influences choice of radio architecture. The CMOS approach to radio design calls for the elimination of discrete components in favor of high levels of on-chip integration which freely use translators and mix analog and digital functionality; in these respects, it departs from traditional RF circuit practices.

20. F. Behbahani et al., "CMOS Mixers and Polyphase Filters for Large Image Rejection," *IEEE Journal of Solid-State Circuits,* June 2001, pp. 873–887.

    This paper presents an in-depth treatment of mixers and polyphase filters and how they are used in rejecting the image in transmitters and receivers. A phasor-based analysis is used to explain all common image-reject topologies and their limitations, and it is shown how this can replace complex trigonometric equations commonly found in the literature. Practical problems in design and layout that limit the performance of image-reject upconversion and down-conversion mixers are identified, and solutions are presented or limits explained. This understanding is put to work in a low IF CMOS wideband, low IF down-conversion circuit, which repeatedly rejects the image by 60 dB over the wide band of 3.5 to 20 MHz without trimming or calibration.

21. R. Carsello, "IMT-2000 Standards: Radio Aspects," paper presented at the IEEE Personal Communications Conference, San Diego, Aug. 1997.

    This article starts by comparing the significant differences between second-generation mobile systems and the major objectives envisioned for IMT-2000, with particular emphasis on the satellite component. It also describes the flexible modular architecture for IMT-2000, which will facilitate the introduction and evolution of new capabilities. The article overviews the technology facilitators that, by the year 2000, will greatly assist the economic implementation of these new systems. Finally, the ITU process for the selection of radio technologies that will fulfil these ambitious requirements is also outlined.

22. J. Crols et al., "Low-IF Topologies for High-Performance Analog Front Ends of Fully Integrated Receivers," *IEEE Transactions on Circuits and Systems—II*, Mar. 1998, pp. 269–282.

    In this paper, the fundamental principles of the low IF receiver topology are introduced. Different low IF receiver topologies are synthesized and fully analyzed in this paper. This is done by applying the complex signal technique—a technique used in digital applications to the study of analog receiver front ends.

23. T. H. Lee, *The Design of CMOS RF ICs*, 2nd Ed., Cambridge University Press, New York, 2004.

    This is a detailed book on the various aspects of the design of CMOS RF ICs. It covers the following: 1. A nonlinear history of radio; 2. Overview of wireless principles; 3. Passive RLC networks; 4. Charateristics of passive IC components; 5. A review of MOS device physics; 6. Distributed systems; 7. The Smith chart and S-parameters; 8. Bandwidth estimation techniques; 9. High frequency amplifier design; 10. Voltage references and biasing; 11. Noise; 12. LNA design; 13. Mixers; 14. Feedback systems; 15. RF power amplifiers; 16. Phase-locked loops; 17. Oscillators and synthesizers; 18. Phase noise; 19. Architectures; 20. RF circuits through the ages.

24. B. Razavi, *RF Microelectronics*, Prentice-Hall, Upper Saddle River, NJ, 1999.

    This is another thorough book on the design of RF ICs and systems. This book begins with a thorough introduction to the fundamental concepts of RF design, including nonlinearity, interference, and noise. It reviews modulation and detection theory, multiple access techniques, and current wireless standards—including CDMA, TDMA, AMPS, and GSM. It presents case studies of transceiver architectures designed by several leading manufacturers. Finally, it offers detailed explanations of low noise amplifiers, mixers, and oscillators; frequency synthesizers; and power amplifiers.

25. R. Adler, "A Study of Locking Phenomena in Oscillators," *Proceedings of the Institute of Radio Engineers*, vol. 34, pp. 351–357, June 1946.

    This is one of the first papers on this topic. Impression of an external signal upon an oscillator of similar fundamental frequency affects both the instantaneous amplitude and instantaneous frequency. Using the assumption that time constants in the oscillator circuit are small compared to the length of one beat cycle, a differential equation is derived which gives the oscillator phase as a function of time. With the aid of this equation, the transient process of "pull-in" as well as the production of a distorted beat note are described in detail. It is shown that the same equation serves to describe the motion of a pendulum suspended in a viscous fluid inside a rotating container. The whole range of locking phenomena is illustrated with the aid of this simple mechanical model.

26. M. Steyaert et al., "A 2V CMOS Cellular Transceiver Front-End," *IEEE Journal of Solid-State Circuits,* Dec. 2000, pp. 1895–1907.

This work presents the design and implementation of a 2-V cellular transceiver front end in a standard 0.25-µm CMOS technology. The prototype integrates a low IF receiver (low noise amplifier, I/Q mixers, and VGAs) and a direct up-conversion transmitter (I/Q mixers and preamplifier) on a single die together with a complete phase-locked loop, including a 64/79 prescaler, a fully integrated loop filter, and a quadrature VCO with on-chip inductors. Design trade-offs have been made over the boundaries of the different building blocks to optimize the overall system performance. All building blocks feature circuit topologies that enable comfortable operation at low voltage. As a result, the IC operates from a power supply of only 2 V, while consuming 191 mW in receiver (RX) mode and 160 mW in transmitter (TX) mode. To build a complete transceiver system for 1,8-GHz cellular communication, only an antenna, an antenna filter, a power amplifier, and a digital baseband chip must be added to the analog front end. This work shows the potential of achieving the analog performance required for the class I/II DCS-1800 cellular system in a standard 0.25-µm CMOS technology without tuning or trimming.

27. J. Fenk, "Highly Integrated RF ICs for GSM and DECT Systems—A Status Review," *JSSC,* Dec. 1997.

TDMA-based digital systems like GSM for cellular and DECT for cordless application have created an increasing market within Europe and gained widespread acceptance also outside Europe. This paper gives an overview of both systems. The system requirements and their influences on highly integrated RF ICs for GSM and DECT are discussed in detail. The various trends of progresses in integration of both systems will be shown, with the different advantages and the disadvantages of the concepts in use. The challenges of increasing the level of integration and an outlook to the future will be presented.

28. J. Strange et al., "A Direct Conversion Transceiver for Multi-Band GSM Application," paper presented at the IEEE RFIC Symposium, Boston, 2000.

The use of direct-conversion receiver topologies has been marked by many technical problems, particularly when used in a TDMA environment. This paper describes a receiver architecture that overcomes many of the traditional problems associated with direct conversion. This architecture has been applied in the design of a GSM multiband transceiver that also features a development of the offset PLL transmitter together with a fast-locking fractional-N synthesizer.

29. A. Behzad et al., "A 5-GHz Direct-Conversion CMOS Transceiver Utilizing Automatic Frequency Control for IEEE 802.11a Wireless LAN Standard," *JSSC,* Dec. 2003.

A fully integrated CMOS direct-conversion 5-GHz transceiver with automatic frequency control is implemented in a 0.18-µm digital CMOS process and housed in an LPCC-48 package. This chip, along with a companion baseband chip, provides a complete 802.11a solution The transceiver consumes 150 mW in receive mode and 380 mW in transmit mode while transmitting +15-dBm output power. The receiver achieves a sensitivity of better than –93.7 and –73.9 dBm for 6 and 54 Mb/s, respectively (even using hard-decision decoding). The transceiver achieves a 4-dB receive noise figure and a +23-dBm transmitter saturated output power. The transmitter also achieves a transmit error vector magnitude of –33 dB. The IC occupies a total die area of 11.7 mm^2 and is packaged in a 48-pin LPCC package. The chip passes better than ±2.5 kV ESD performance. Various integrated self-contained or system level calibration capabilities allow for high performance and high yield.

30. D. H. Morais and K. Feher, "The Effects of Filtering and Limiting on the Performance of QPSK, Offset QPSK, and MSK Signals," *IEEE Transactions on Communications,* vol. 28, pp. 1999–2006, Dec. 1980.

The effects on spectral densities, symbol wave shapes, and Pe versus S/N performances, resulting from the addition of filtering followed by hard limiting on raised cosine filtered QPSK, offset QPSK, and MSK point-to-point radio systems, are studied. A mathematical model and physical insight are presented into the crosstalk phenomenon between quadrature channels, created in the systems by the effect of limiting on filtered signals. This crosstalk is shown to result whether the filtering is ideal or otherwise. Computer generated and measured eye diagrams showing crosstalk as predicted on a filtered, then limited, offset QPSK signal are given. Measured and computed spectral density results are given which are in close agreement with each other, indicating that the computer model provides a good representation of the real system. In addition, an explanation of the shape of the power spectra associated with filtered, then limited, modulated signals is provided by studying the symbol wave shapes of these signals. Using the spectral density results and the Pe (S/N) performance findings, it is shown that for a microwave system which (1) incorporates an amplitude-limiting amplifying device in the transmitter, (2) must operate within the FCC limits for radiated spectrum, and (3) must operate at a spectral efficiency greater than 1 bit/s/Hz, offset QPSK modulation is the best choice of the three modulation methods studied.

31. J. F. Sevic and J. Staudinger, "Simulations of Adjacent Channel Power for Digital Wireless Communication Systems," *Microwave Journal,* pp. 66–80, Oct. 1996.

A comparison of nonlinear analysis methods for simulation of power amplifier adjacent-channel power ratio is presented. Adjacent-channel power ratio is the linearity figure of merit for digital wireless communication systems employing nonconstant-envelope modulation techniques, such as OQPSK and $\pi/4$-QPSK. Trade-offs in the performance of each method are discussed. Using modulation for the TIA IS-95 and the TIA IS-54 standards, measured and simulated results for a single-stage power amplifier are presented

32. A. Hajimiri and T. Lee, " Design Issues in CMOS Differential LC Oscillators," *JSSC,* pp. 717–724, May 1999.

An analysis of phase noise in differential cross-coupled inductance–capacitance (LC) oscillators is presented. The effect of tail current and tank power dissipation on the voltage amplitude is shown. Various noise sources in the complementary cross-coupled pair are identified, and their effect on phase noise is analyzed. The predictions are in good agreement with measurements over a large range of tail currents and supply voltages. A 1.8-GHz LC oscillator with a phase noise of -121 dBc/Hz at 600 kHz is demonstrated, dissipating 6 mW of power using on-chip spiral inductors

33. P. Gray, P. Hurst, S. Lewis, and R. Meyer, *Analysis and Design of Analog Integrated Circuits,* 4th ed., Wiley, New York, 2001.

This edition of the book features coverage of several topics—more advanced CMOS device electronics to include short-channel effects, weak inversion, and impact ionization. In addition, coverage of state-of-the-art IC processes shows how modern integrated circuits are fabricated, including heterojunction bipolar transistors, copper interconnect, and low permittivity dielectric materials. A comprehensive and unified treatment of bipolar and CMOS circuits is presented.

34. T. H. Lee, *Planar Microwave Engineering,* Cambridge University Press, New York, 2004.

This book covers many practical techniques for microwave design and measurements. The book covers the following: 1. A microhistory of microwave technology; 2. Introduction to RF and microwave technology; 3. The Smith chart and S-parameters; 4. Impedance matching; 5. Connectors, cables, and waveguide; 6. Lumped passive components; 7. Microstrip, stripline, and planar passive elements; 8. Impedance measurement; 9. Microwave diodes; 10. Mixers; 11. Transistors; 12. Small-signal amplifiers; 13. Low noise amplifiers; 14. Noise figure measurement; 15. Oscillators; 16. Synthesizers; 17. Oscillator phase noise; 18. Phase noise measurement; 19. Sampling oscilloscopes, spectrum analyzers, and probes; 20. Power amplifiers; 21. Antennas; 22. Lumped filters; 23. Microstrip filters.

35. J. G. Proakis and M. Salehi, *Communication Systems Engineering*, Prentice-Hall, Upper Saddle River, NJ, 2002.

With an emphasis on digital communications, this edition of the book introduces the basic principles underlying the analysis and design of communication systems. In addition, this text gives a solid introduction to analog communications and a review of important mathematical foundation topics.

36. S. Mehta et al., "An 802.11g WLAN SoC," *IEEE International Solid-State Circuits Conference Digest of Technical Papers*, pp. 94–95, Feb. 2005.

A single-chip IEEE-802.11g-compliant wireless LAN system-on-a-chip (SoC) that implements all RF, analog, digital PHY, and MAC functions has been integrated in a 0.18-μm CMOS technology. The IC transmits 0-dBm EVM-compliant output power for a 64-QAM OFDM signal. The overall receiver sensitivities are better than –92 and –73 dBm for data rates of 6 and 54Mb/s, respectively.

37. D. Tse and P. Viswanath, *Fundamentals of Wireless Communications*, Cambridge University Press, New York, 2005.

This textbook takes a unified view of the fundamentals of wireless communication and explains the web of concepts underpinning these advances at a level accessible to an audience with a basic background in probability and digital communication. Topics covered include MIMO (multiple input, multiple output) communication, space–time coding, opportunistic communication, OFDM, and CDMA. The concepts are illustrated using many examples from wireless systems such as GSM, IS-95 (CDMA), IS-856(1xEV-DO), Flash OFDM, and ArrayComm SDMA systems. Particular emphasis is placed on the interplay between concepts and their implementation in systems. An abundant supply of exercises and figures reinforce the material in the text. This book is intended for use on graduate courses in electrical and computer engineering and will also be of great interest to practicing engineers.

38. A. Behzad et al. "A Fully Integrated MIMO Multiband Direct Conversion CMOS Transceiver for WLAN Apllications (802.11n)," *IEEE International Solid-State Circuits Conference Digest of Technical Papers*, pp. 560–561, Feb. 2007.

A single-chip, multiband, direct-conversion CMOS MIMO transceiver (2×2) targeted for WLAN applications is presented. This transceiver is capable of satisfying the requirements of the draft 802.11n standard and achieves PHY rates of >270 Mbps. The receivers and transmitters achieve an EVM of better than –41 dB (0.9%) and –40 dB (1.0%) operating in legacy g and a modes, respectively. From a 1.8-V supply and with both cores operating, the chip draws 275 mA in RX mode and 280 mA in TX mode.

39. D. Rahn et al., "A Fully Integrated Multiband MIMO WLAN Transceiver RFIC," *JSSC*, pp. 1629–1641, Aug. 2005.

A multiple input–multiple output (MIMO) transceiver RFIC compliant with IEEE

802.11a/b/g and Japan wireless LAN (WLAN) standards is presented. The transceiver has two complete radio paths integrated on the same chip. When two chips are used in tandem to form a four-path composite beam forming (CBF) system, 15 dB of link margin improvement is obtained. The transceiver was implemented in a 47-GHz SiGe technology with 29.1 mm² die size. It consumes 195 mA in RX mode and 240 mA in TX mode from a 2.75-V supply.

40. G. Chien et al., "A Fully-Integrated Dual-Band MIMO Transceiver IC," *RFIC Digest of Technical Papers,* 4 pp., 2006.

    A monolithic MIMO transceiver IC consisting of two transmitters and three receivers is implemented in a 0.35/spl mu/m SiGe BiCMOS process. The receivers achieve a NF of 4 dB in 2.4 GHz and 5.5 dB in 5 GHz, while the transmitters deliver an OP1dB of 11 dBm. The MIMO transceiver in full operation consumes approximately 260 mA in RX mode and 245 mA in TX mode from a 3-V supply.

41. Y. Palaskas et al., "A 5GHz 108Mb/s 2×2 MIMO Transceiver with Fully Integrated +16dBm PAs in 90nm CMOS," *IEEE International Solid-State Circuits Conference Digest of Technical Papers,* pp. 368–369, Feb. 2006.

    This paper presents a fully integrated 5-GHz 2 × 2 MIMO WLAN transceiver RFIC implemented in 90-nm CMOS. The paper identifies the key MIMO integration issues and proposes techniques to optimize MIMO performance. It is shown that crosstalk between the multiple transceivers residing on the same die can degrade MIMO performance and has to be carefully minimized, especially when power amplifiers are integrated on die. A shared LO generation and distribution network is designed to maximize MIMO phase noise immunity without introducing undesired crosstalk. The fabricated MIMO receiver achieves a sensitivity of –63 dBm while receiving 108 Mbps in MIMO spatial multiplexing mode in the presence of a 25-ns Rayleigh fading channel. The sensitivity of a single receiver in the presence of AWGN noise is –76 dBm. Linearized 3.3-V, 5-GHz power amplifiers with P1dB = 20.5 dBm deliver average power of +13 and +16 dBm each in MIMO and SISO modes, respectively (EVM = –27/–25 dB). The measured performance demonstrates the effectiveness of the isolation techniques employed. The system in a package includes an 18-mm² die and microstrip front-end matching networks implemented on a flip-chip package.

42. C. P. Lee et al., "A Highly Linear Direct-Conversion Transmit Mixer Transconductance Stage with Local Oscillation Feedthrough and I/Q Imbalance Cancellation Scheme," *IEEE International Solid-State Circuits Conference Digest of Technical Papers,* pp. 368–369, Feb. 2006.

    Some of the requirements of the next generation WLAN transmitters are low transmit EVM, a low local oscillator feedthrough (LOFT), a small I/Q imbalance, a wide gain-control range, and preferably a minimum number of real-time calibrations. In this paper, a highly linear transmit mixer transconductance stage is presented that incorporates a wide gain-control range and a one-time LOFT and I/Q imbalance cancellation scheme to meet these goals.

43. K. Bult et al., "An Inherently Linear and Compact MOST-Only Current Division Technique," *IEEE Journal of Solid-State Circuits,* vol. 27, no. 12, pp. 1730–1735, Dec. 1992.

    A technique is presented for dividing currents accurately and linearly by using MOS transistors only. This technique is valid in all operating regions of an MOS transistor. With this technique, a volume control circuit is realized with an attenuation of 0 to –84 dB in steps of 2 dB. The measured THD is better than –85 dB and the dynamic range is better than 100 dB. The chip is realized in a standard digital CMOS process and chip area is 0.22 mm².

44. A. Behzad et al., "A 4.92–5.845 GHz Direct-Conversion CMOS Transceiver for IEEE 802.11a Wireless LAN," *RFIC Digest of Technical Papers*, 2004, pp. 335–338.

   A fully integrated CMOS direct-conversion 5-GHz transceiver is implemented in a 0.18-μm digital CMOS process and housed in an LPCC-48 package. This chip, along with a companion baseband chip, provides a complete 802.11a solution covering all of the worldwide 4.92- to 5.845-GHz bands. The receiver achieves a 3.5-dB NF while the transmitter achieves a +23-dBm saturated output power. The integrated PA utilizes a linearization technique to allow for high efficiency while maintaining the linear operation required by QAM64 OFDM signals. The transceiver achieves low cost and high yield through the use of various integrated self-contained or system level calibration techniques.

45. A. Behzad, "Radio Design for MIMO Systems with an Emphasis on IEEE 802.11n," course presented at IEEE International Solid-State Circuits Conference, San Francisco, 2007.

   Essential to the overall system design of a MIMO system is the radio design. This course provides a brief introduction to the legacy 802.11 a/b/g systems, followed by a discussion of the history of multiple antenna systems and the conventional analog-based techniques such as MRC. A general introduction to the 802.11n then follows, which includes the channelization and modulation types, the definition and the description of the concepts behind the multiple spatial streams (M × N), and additional PHY and MAC techniques allowing for higher rates and/or longer reach. These features include the use of short guard interval (GI), implicit and explicit beamforming, space–time block codes (STBC), the use of Greenfield mode, and aggregation techniques. The requirements of the 802.11n standard such as sensitivity and EVM and their relation to analog impairments such as phase noise, quadrature imbalances, linearity, and crosstalk are also discussed. Some specific circuit examples are presented and some unique circuit implementation challenges of MIMO radios are discussed. Some measured performance numbers (range and throughput) will be also presented. The course wraps up by discussing the future trends of MIMO radio implementation.

46. C. Balanis, *Antenna Theory*, 3rd ed., Wiley, Hoboken, NJ, 2005.

   The discipline of antenna theory has experienced vast technological changes. In response, the author has updated his classic text, offering a recent look at the necessary topics. New material includes smart antennas and fractal antennas, along with the latest applications in wireless communications. Multimedia material on an accompanying CD presents PowerPoint viewgraphs of lecture notes, interactive review questions, Java animations and applets, and MATLAB features. Like the previous editions, this third edition is appropriate for electrical engineering and physics students at the senior undergraduate and beginning graduate levels and practicing engineers as well.

47. R. J. Baker, *CMOS Circuit Design, Layout and Simulation*, 2nd ed., Wiley/IEEE, Hoboken, NJ, 2005

   This book covers the practical design of both analog and digital integrated circuits, offering a contemporary view of a wide range of analog/digital circuit blocks, the BSIM model, data converter architectures, and other topics. This edition takes a two-path approach to the topics; design techniques are developed for both long- and short-channel CMOS technologies and then compared. The results are multidimensional explanations that allow readers insight into the design process.

48. R. J. Baker, *CMOS Mixed-Signal Circuit Design*, Wiley/IEEE, Hoboken, NJ, 2002.

   This book builds on the fundamental material in the author's previous book, CMOS: Circuit Design, Layout, and Simulation, *to provide a textbook and reference for mixed-signal circuit design. The coverage is both practical and in-depth, integrating experimental, theoretical, and simulation examples to drive home the why and the how of doing mixed-signal circuit design. Some of the highlights of this book include a practical/theoretical approach to mixed-signal circuit design with an emphasis on oversampling techniques; coverage of delta–sigma data converters, custom analog and digital filter design, design with submicrometer CMOS processes, and practical debug prototyping techniques.*

Index

References

Adler, R. (1946), "A Study of Locking Phenomena in Oscillators," *Proc. I.R.E. Waves Electrons,* pp. 351–357.

Behzad, A., et al. (2003, Feb.). "A Direct-Conversion CMOS Transceiver with Automatic Frequency Control for IEEE 802.11a Wireless LAN," *IEEE ISSCC Dig. Tech. Papers,* pp. 356–499.

Behzad, A., et al. (2004a), "A 4.92–5.845 GHz Direct-Conversion CMOS Transceiver for IEEE 802.11a Wireless LAN," *IEEE RFIC Dig. Tech. Papers,* pp. 335–338.

Behzad, A., et al., (2004b, Dec.), "A 5 GHz Direct-Conversion CMOS Transceiver Utilizing Automatic Frequency Control for IEEE 802.11a Wireless LAN Standard," *IEEE JSSC,* pp. 2209–2220.

Behzad, A., et al. (2007a, Feb.), "A Fully Integrated MIMO Multiband Direct Conversion CMOS Transceiver for WLAN Apllications (802.11n)," *IEEE ISSCC Dig. Tech. Papers,* pp. 560–561.

Behzad, A., et al. (2007b, Dec.), "An 802.11a/b/g/n Direct Conversion MIMO Transceiver in Digital CMOS," *IEEE JSSC,* to be published.

Bouras, I., et al. (2003, Dec.), "A Single-Chip Digitally Calibrated 5.15-5.825-GHz 0.18μm CMOS Transceiver for 802.11a Wireless Lan," *IEEE JSSC,* pp. 2221–2231.

Bult, K., et al. (1992, Dec.), "An Inherently Linear and Compact MOST-Only Current Division Technique," *IEEE J. Solid-State Circuits,* vol. 27, no. 12, pp. 1730–1735.

Cavers, J. (1997, Aug.), "New Methods for Adaptation of Quadrature Modulators and Demodulators in Amplifier Linearization Circuits," *IEEE Trans. Vehic. Technol.,* vol. 46, no. 3, pp. 707–716.

Chien, G., et al. (2006, Jun.), "A Fully-Integrated Dual-Band MIMO Transceiver IC," *IEEE RFIC Dig. Tech. Papers,* 4 pp.

Cutler, B. (2002, May), "Effects of Physical Layer Impairments on OFDM Systems," *RF Design.*

Darabi, H., et al. (2001, Dec.), "A 2.4-GHz CMOS Transceiver for Bluetooth," *IEEE JSCC,* pp. 2016–2024.

Gray, P. and Mayer, R. (1984). *Analysis and Design of Analog Integrated Circuits,* Second Edition, Wiley.

Hajimiri, A. and Lee T. H. (1998, Feb.), "A General Theory of Phase Noise in Electrical Oscillators," *IEEE JSSC,* pp. 179–194.

Kim, J., et al. (2005). "Prediction of Error Vector Magnitude Using AM/AM, AM/PM Distortion of RF Power Amplifier for High Order Modulation OFDM System," *Microwave Symp. Dig.,* IEEE MTT-S International.

Lee, T. H. (1998), *The Design of CMOS Radio Frequency Circuits,* Cambridge University Press, New York.

Lee, C. P., et al. (2006, Feb.), "A Highly Linear Direct-Conversion Transmit Mixer Transconductance Stage with Local Oscillation Feedthrough and I/Q Imbalance Cancellation Scheme," *IEEE ISSCC Dig. Tech. Papers,* pp. 368–369.

Mehr, I., et al. (1997, Apr.), "A CMOS Continuous-Time Gm-C Filter for PRML Read Channel Applications at 150 Mb/s and Beyond," *IEEE J. Solid-State Circuits,* vol. 32, no. 4, pp. 499–513.

Palaskas, Y., et al. (2006, Feb.), "A 5GHz 108Mb/s 2×2 MIMO Transceiver with Fully Integrated +16dBm PAs in 90nm CMOS," *IEEE ISSCC Dig. Tech. Papers,* pp. 368–369.

Rahn, D., et al. (2005, Aug.), "A Fully Integrated Multiband MIMO WLAN Transceiver RFIC," *IEEE JSSC,* pp. 1629–1641.

Rappaport, T. S. (1996), *Wireless Communications—Principles & Practice,* IEEE Press, New York.

Razavi, B. (1998), *RF Microelectronics,* Prentice-Hall, Englewood Cliffs, NJ.

Rofougaran, A. R., et al. (2005, Mar.), *IEEE Commun. Mag.*

Sevic, J., et al. (1996, Oct.), "Simulations of Adjacent Channel Power for Digital Wireless Communication Systems," *Microwave J.*

Su, D., et al. (2002), "A 5GHz CMOS Transceiver for 802.11a Wireless LAN," *IEEE ISSCC Dig. Tech. Papers,* .

Tse, D., et al. (2005), *Fundamentals of Wireless Communications,* Cambridge University Press, New York.

Zagari, M., et al. (2002, Dec.), "A 5-GHz CMOS Transister for IEEE 802.11a Wireless LAN Systems," *IEEE JSSC,* pp. 1688–1694.

Printed and bound by CPI Group (UK) Ltd, Croydon, CR0 4YY

16/04/2025

14658606-0001